T0345325

An Introduction to

Numerical Computation

An Introduction to
Numerical Computation

Wen Shen

Penn State University, USA

NEW JERSEY · LONDON · SINGAPORE · BEIJING · SHANGHAI · HONG KONG · TAIPEI · CHENNAI · TOKYO

Published by

World Scientific Publishing Co. Pte. Ltd.

5 Toh Tuck Link, Singapore 596224

USA office: 27 Warren Street, Suite 401-402, Hackensack, NJ 07601

UK office: 57 Shelton Street, Covent Garden, London WC2H 9HE

Library of Congress Cataloging-in-Publication Data

Names: Shen, Wen, 1968–

Title: An introduction to numerical computation / Wen Shen (Penn State University, USA).

Description: New Jersey : World Scientific, 2016. |

   Includes bibliographical references and index.

Identifiers: LCCN 2015033650 | ISBN 9789814730068 (hardcover : alk. paper)

Subjects: LCSH: Numerical analysis--Textbooks.

Classification: LCC QA297 .S466 2016 | DDC 518--dc23

LC record available at http://lccn.loc.gov/2015033650

British Library Cataloguing-in-Publication Data

A catalogue record for this book is available from the British Library.

Printed in Singapore

To my parents,
Zhi-Lian Xuan and Qing-Jiang Shen

献给我的父母亲，
宣志莲和沈情江。

Preface

This is a set of lecture notes developed during a span of 10 years, for an introductory course on numerical computation which I have been teaching to senior undergraduate students at Penn State University. The course is cross-listed between the two departments of Mathematics and Computer Science. It is a standard 3-credit course, meeting 3 times per week, for a total of 15 weeks.

These lecture notes contain precisely the material that can be covered in such a course, together with a few optional sections which may provide additional reading. The notes are written in a rather colloquial style, presenting the subject matter in the same form as it can be explained in a classroom. For instructors, this will minimize the amount of effort required to prepare their blackboard presentations. For students, these notes can also be used as a companion to the videos posted on my Youtube channel at "youtube.com/wenshenpsu". These videos contain a complete set of live lectures for the course, which I recorded during the 2015 Spring semester. Shorter tutorials, lasting 5 to 15 minutes each and covering specific topics or examples, are also freely available on my Youtube channel.

As prerequisites, students should have taken the standard calculus courses (in one and in several variables), an introductory course on matrices, and a course on computer programming. Having attended a sophomore-level course on differential equations is desirable, but not strictly required.

After an introductory chapter on computer arithmetic, Chapters 2–4 of these notes cover polynomial approximations and their use for numerical integration. The following Chapters 5–7 are focused on the numerical solution of nonlinear equations and of linear systems. Data fitting, using least squares methods, is discussed in Chapter 8. The remainder of the book is concerned with the numerical solution of differential equations. Initial value problems for ODEs are covered in Chapter 9, while Chapter 10 describes some basic methods to solve two-point boundary value problems. Finally, the last chapter provides a brief introduction to finite difference schemes for partial differential equations. This is achieved by means of a few elementary examples, where the Laplace equation and the heat equations are solved on a rectangular grid, and the resulting systems of linear equations are discussed.

For each class of problems, I decided to present only a few, basic computational methods. Rather than surveying a large number of algorithms, I emphasize the key mathematical ideas that motivate these methods. In most cases, rigorous proofs are skipped, relying instead on graphs and drawings in order to build up intuition. Still, some simple proofs are fully worked out, introducing students to the elegance of mathematical reasoning.

Implementation of the algorithms is an important part of the course. Although various program languages could be used, and many software packages are suitable for scientific computing, I here prefer to use Matlab. Tutorials on Matlab can be found online. Many Matlab examples and codes are included in this lecture notes, as a further help to students.

A set of homework problems is included at the end of each chapter. These assignments typically include some simple calculations done with pencil and paper, together with larger computational projects, using Matlab. Lab projects cover a variety of applications, in connection with population dynamics, engineering, image reconstruction, mechanics, etc. They are designed to illustrate the wide range of applications and the power of numerical algorithms. Students in my class are required to complete every problem in the homework sets.

In my experience, for many students this is a challenging course, requiring a combination of knowledge and skills of various kind. Students taking the class have different backgrounds, some of them majoring in mathematics, physics, engineering, computer science, etc. Some students are very fluent in computer coding, but may feel intimidated when confronted with an abstract proof. Others find it easy to understand mathematical ideas, but will need help with Matlab coding. It is this variety of backgrounds that makes it particularly hard to teach a course on numerical computation. Lectures must be designed so that every student receives the help he/she needs, and is motivated to learn new material. In the course of several years, these lecture notes have received very positive reviews from Penn State students. By publishing them as a book, and in conjunction with the Youtube videos, I hope a wider audience will also find them useful.

I am grateful to my husband Alberto Bressan for carefully reading the draft, finding numerous errors, and making many suggestions for substantial improvement.

Wen Shen
Summer 2015
State College, Pennsylvania, USA

Contents

Chapter 1

Computer Arithmetic

1.1 Introduction

What are numerical methods? They are algorithms that compute approximate solutions to a number of problems for which exact solutions are not available. These include, but are not limited to:

- evaluating functions,
- computing the derivative of a function,
- computing the integral of a function,
- solving a nonlinear equation,
- solving a system of linear equations,
- finding solutions to an ODE (ordinary differential equation),
- finding solutions to a PDE (partial differential equation).

Such algorithms can be implemented (programmed) on a computer.

The following diagram shows how various aspects are related.

Keep in mind that a course on numerical methods is NOT about numbers. It is about mathematical ideas and insights.

We shall study some basic issues:

- development of algorithms;
- implementation;
- a little bit of analysis, including error-estimates, convergence, stability etc.

Throughout the course we shall use <u>Matlab</u> for programming purposes.

1.2 Representation of Numbers in Different Bases

Historically, there have been several bases for representing numbers, used by different civilizations. These include:

  10: decimal, which has become standard nowadays;
   2: binary, computer use;

1

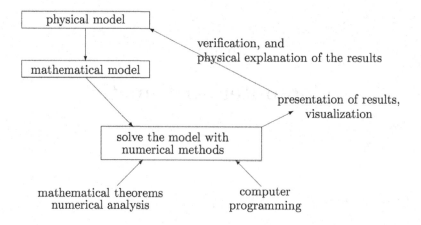

Fig. 1.1   The big picture

8: octal;

16: hexadecimal, used in ancient China;

20: vigesimal, used in ancient France (numbers 70 to 79 are counted as 60+10 to 60+19 in French, and 80 is 4 × 20);

60: sexagesimal, used by the Babylonians.

In principle, one can use any integer β as the base. The value of a number in base β is then written as

$$
\left(\overbrace{a_n a_{n-1} \cdots a_1 a_0}^{\text{integer part}} . \overbrace{b_1 b_2 b_3 \cdots}^{\text{fractional part}} \right)_\beta
$$

$$
= a_n \beta^n + a_{n-1} \beta^{n-1} + \cdots + a_1 \beta + a_0 \qquad \text{(integer part)}
$$
$$
+ b_1 \beta^{-1} + b_2 \beta^{-2} + b_3 \beta^{-3} + \cdots \qquad \text{(fractonal part)}
$$

Taking $\beta = 10$, we obtain the standard decimal representation.

The above formula allows us to convert a number in any base β into decimal base.

In principle, one can convert the numbers between different bases. Here are some examples.

Example 1.1. Conversion of octal $\rightarrow$ decimal:

$$
(45.12)_8 = 4 \times 8^1 + 5 \times 8^0 + 1 \times 8^{-1} + 2 \times 8^{-2} = (37.15625)_{10}
$$

Example 1.2. Conversion of octal $\leftrightarrow$ binary:

Observe that, since $8 = 2^3$, we have

$$(1)_8 = (001)_2$$
$$(2)_8 = (010)_2$$
$$(3)_8 = (011)_2$$
$$(4)_8 = (100)_2$$
$$(5)_8 = (101)_2$$
$$(6)_8 = (110)_2$$
$$(7)_8 = (111)_2$$
$$(10)_8 = (1000)_2$$

Then, the conversion between these two bases become much simpler. To convert from octal to binary, we can convert each digit in the octal base, and write it into binary base, using 3 digits in binary for each digit in octal. For example:

$$(5034)_8 = (\underbrace{101}_{5} \underbrace{000}_{0} \underbrace{011}_{3} \underbrace{100}_{4})_2$$

To convert a binary number to an octal one, we can group the binary digit in groups of 3, and write out the octal base value for each group. For example:

$$(110\,010\,111\,001)_2 = (\underbrace{6}_{110} \underbrace{2}_{010} \underbrace{7}_{111} \underbrace{1}_{001})_8$$

Example 1.3. Conversion of decimal $\to$ binary: write $(12.45)_{10}$ in binary base.

This example is of particular interest. Since the computer uses binary base, how would the number $(12.45)_{10}$ look like in binary base? How does a computer store this number in its memory? The conversion takes two steps.

First, we treat the integer part. The procedure is to keep dividing by 2, and store the remainders of each step, until one cannot divide anymore. Then, collect the remainder terms in the reversed order. We have

$$
\begin{array}{r|ll}
 & (remainder) & \\
2 & \underline{12} & 0 & \\
2 & \underline{6} & 0 & \Rightarrow (12)_{10} = (1100)_2. \\
2 & \underline{3} & 1 & \\
2 & \underline{1} & 1 & \\
\end{array}
$$

Could you figure out a simple proof for this procedure?

For the fractional part, we will keep multiplying the fractional part by 2, and store the integer part for the result. This gives us:

$$
\begin{array}{l|l}
0.45 & \times\,2 \\
\underline{0}.9 & \times\,2 \\
\underline{1}.8 & \times\,2 \\
\underline{1}.6 & \times\,2 \\
\underline{1}.2 & \times\,2 \\
\underline{0}.4 & \times\,2 \\
\underline{0}.8 & \times\,2 \\
\underline{1}.6 & \times\,2 \\
\cdots &
\end{array}
\qquad \Rightarrow (0.45)_{10} = (0.01110011001100\cdots)_2.
$$

Putting them together, we get

$$(12.45)_{10} = (1100.01110011001100\cdots)_2.$$

Again, would you be able to prove why this procedure works?

It is surprising that a simple finite length decimal number such as 12.45 corresponds to a fractional number with infinite length, when written in binary form! This is very bad news! To store this number *exactly* in a computer, one would need infinite memory space!

How does a computer store such a number? Below we will learn the standard representation used in modern computers.

1.3   Floating Point Representation

We recall the normalized scientific notation for a real number x:

Decimal: $x = \pm r \times 10^n$, $1 \le r < 10$ (Example: $2345 = 2.345 \times 10^3$)

Binary: $x = \pm r \times 2^n$, $1 \le r < 2$

Octal: $x = \pm r \times 8^n$, $1 \le r < 8$ (just as an example)

For any base β: $x = \pm r \times \beta^n$, $1 \le r < \beta$

We observe that the important information to be stored include: (i) the sign, (ii) the exponent n, and (iii) the value of r.

The computer uses the binary version of the number system. It represents numbers with finite length. This means: it either rounds off or chops off after a certain fractional point. These are called *machine numbers*.

The value r is called *the normalized mantissa*. For binary numbers, r lies between 1 and 2, so we have

$$r = 1.(\text{fractional part})$$

Therefore, to save space, in the computer we will only store the fractional part of the number.

The integer number n is called the exponent. If $n > 0$, then we have $x > 1$. If $n < 0$, then $x < 1$.

Each bit in a computer can store the value either 0 or 1.

In a 32-bit computer, with single-precision, a number is stored as follows:

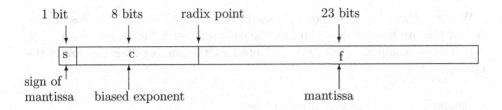

Fig. 1.2   32-bit computer with single precision

Note that the value c, which takes 8-bit space, is not the actual exponent, but the *biased exponent*. The maximum number of values that can be stored in the space of 8-bit is $2^8 = 256$. Since we want to represent both positive and negative exponents in an equal way, we will use the space to represent numbers from -127 to 128. The actual value of the exponent is thus $n = c - 127$.

Therefore, the actual value of the number in Figure 1.2 in the computer is:

$$(-1)^s \times 2^{c-127} \times (1.f)_2.$$

Here $r = (1.f)_2$ is the value of the binary number, where f is the fractional part, which is stored in 23 bits.

This representation of numbers is called: *single-precision IEEE standard floating-point*.

The smallest representable number in absolute value is obtained when the exponent is smallest, i.e., $n = -127$, and we have

$$x_{\min} = 2^{-127} \approx 5.9 \times 10^{-39}.$$

The largest representable number in absolute value is obtained when the exponent is largest, i.e., $n = 128$, and we have

$$x_{\max} = 2^{128} \approx 2.4 \times 10^{38}.$$

It is clear that computers can only handle numbers with absolute values between $x_{\min}$ and $x_{\max}$.

We say that x *underflows* if $|x| < x_{\min}$. In this case, we consider $x = 0$.

We say that x *overflows* if $|x| > x_{\max}$. In this case, we consider $x = \infty$.

Error in the floating point representation. Let $\mathrm{fl}(x)$ denote the floating point representation of the number x. In general it contains error, caused by roundoff or chopping. Let δ denote the relative error. We have

$$\mathrm{fl}(x) = x \cdot (1 + \delta)$$

$$\text{relative error: } \delta = \frac{\text{fl}(x) - x}{x}$$

$$\text{absolute error: } = \text{fl}(x) - x = \delta \cdot x$$

We know that $|\delta| \leq \varepsilon$, where ε is called *machine epsilon*, which represents the smallest positive number detectable by the computer, such that $\text{fl}(1 + \varepsilon) > 1$.

In a 32-bit computer: $\varepsilon = 2^{-23}$. (Note that the number 23 is the size of the mantissa.)

Computer errors in representing numbers:

- relative error in rounding off: $\delta \leq 0.5 \times 2^{-23} \approx 0.6 \times 10^{-7}$
- relative error in chopping: $\delta \leq 1 \times 2^{-23} \approx 1.2 \times 10^{-7}$

Error propagation (through arithmetic operation). When a computer performs basic arithmetic operations such as addition or multiplication, the error in the representation of the numbers will be carried on, and appear in the result. One can image that, after a series of computations, the errors will be carried on at each step, and possibly accumulate over time. This is called *error propagation*. We demonstrate this fact by an example of addition.

Example 1.4. Consider an addition, say $z = x + y$, done in a computer. How would the errors be propagated?

To fix the idea, let $x > 0, y > 0$, and let $\text{fl}(x), \text{fl}(y)$ be their floating point representation. Then, we have

$$\text{fl}(x) = x(1 + \delta_x), \qquad \text{fl}(y) = y(1 + \delta_y)$$

where δ_x and δ_y are the errors in the floating point representation for x and y, respectively. Then

$$\begin{aligned}
\text{fl}(z) &= \text{fl}\left(\text{fl}(x) + \text{fl}(y)\right) \\
&= \left(x(1 + \delta_x) + y(1 + \delta_y)\right)(1 + \delta_z) \\
&= (x + y) + x \cdot (\delta_x + \delta_z) + y \cdot (\delta_y + \delta_z) + (x\delta_x\delta_z + y\delta_y\delta_z) \\
&\approx (x + y) + x \cdot (\delta_x + \delta_z) + y \cdot (\delta_y + \delta_z)
\end{aligned}$$

Here, δ_z is the round-off error in making the floating point representation for z. Then, we have

$$\text{absolute error} = \text{fl}(z) - (x + y) = x \cdot (\delta_x + \delta_z) + y \cdot (\delta_y + \delta_z)$$

$$= \underbrace{x \cdot \delta_x}_{\substack{\text{abs. error} \\ \text{for } x}} + \underbrace{y \cdot \delta_y}_{\substack{\text{abs. error} \\ \text{for } y}} + \underbrace{(x + y) \cdot \delta_z}_{\text{round-off error}}$$

$$\underbrace{}_{\text{propagated error}}$$

and
$$\text{relative error} = \frac{\text{fl}(z) - (x + y)}{x + y} = \underbrace{\frac{x\delta_x + y\delta_y}{x + y}}_{\text{propagated error}} + \underbrace{\delta_z}_{\text{round-off error}}$$

Would you be able to work out the error propagation for other arithmetic operations?

1.4 Loss of Significance

Loss of significance typically happens when we compute the difference between two numbers very close to each other. The result will then contain fewer significant digits.

To understand this fact, consider a number with 8 significant digits:
$$x = 0.d_1 d_2 d_3 \cdots d_8 \times 10^{-a}$$
d_1 is the most significant digit, and d_8 is the least significant digit.

Let $y = 0.b_1 b_2 b_3 \cdots b_8 \times 10^{-a}$. We want to compute $x - y$.

Assume that $x \approx y$, such that $b_1 = d_1$, $b_2 = d_2$, $b_3 = d_3$. Then we have
$$x - y = 0.000c_4 c_5 c_6 c_7 c_8 \times 10^{-a}.$$

We have only 5 significant digits in the answer. We lost 3 significant digits. This is called *loss of significance*, which will result in loss of accuracy. The number $x - y$ has a much larger relative error in its representation, and therefore is not so accurate anymore. In designing a numerical algorithm, one should always be sensitive to such a loss, and avoid such subtractions by performing other equivalent computations. We will see some examples below.

Example 1.5. Find the roots of $x^2 - 40x + 2 = 0$. Use 4 significant digits in the computation.

Answer. The roots for the equation $ax^2 + bx + c = 0$ are
$$r_{1,2} = \frac{1}{2a}\left(-b \pm \sqrt{b^2 - 4ac}\right)$$
In our case, we have
$$x_{1,2} = 20 \pm \sqrt{398} \approx 20.00 \pm 19.95$$
therefore
$$x_1 \approx 20 + 19.95 = 39.95 \quad \text{(good)},$$
$$x_2 \approx 20 - 19.95 = 0.05 \quad \text{(bad, since we lost 3 significant digits)}.$$

To avoid such a loss, we can change the algorithm. Observe that $x_1 x_2 = c/a$. Then x_2 can also be computed through a division instead:
$$x_2 = \frac{c}{ax_1} = \frac{2}{1 \cdot 39.95} \approx 0.05006$$
In this way, we get back 4 significant digits in the result.

Example 1.6. We want to numerically compute the function

$$f(x) = \frac{1}{\sqrt{x^2 + 2x} - x - 1}.$$

Explain what problem you might run into for certain values of x. Find a way to overcome this difficulty.

Answer. We see that, for large values of x with $x > 0$, the values $\sqrt{x^2 + 2x}$ and $x + 1$ are very close to each other. Therefore, in the subtraction we will lose many significant digits. To avoid this problem, we rewrite the function $f(x)$ in an equivalent form that does not require any subtraction. This can be achieved by multiplying both numerator and denominator by the conjugate of the denominator, namely

$$f(x) = \frac{\sqrt{x^2 + 2x} + x + 1}{\left(\sqrt{x^2 + 2x} - x - 1\right)\left(\sqrt{x^2 + 2x} + x + 1\right)}$$

$$= \frac{\sqrt{x^2 + 2x} + x + 1}{x^2 + 2x - (x + 1)^2} = -\left(\sqrt{x^2 + 2x} + x + 1\right).$$

1.5   Review of Taylor Series and Taylor Theorem

Assume that $f(x)$ is smooth function, so that all its derivatives exist. Then the Taylor expansion of f about the point $x = c$ is:

$$f(x) = f(c) + f'(c)(x - c) + \frac{1}{2!}f''(c)(x - c)^2 + \frac{1}{3!}f'''(c)(x - c)^3 + \cdots. \quad (1.5.1)$$

Using the summation sign, it can be written in the more compact way

$$f(x) = \sum_{k=0}^{\infty} \frac{1}{k!}f^{(k)}(c)(x - c)^k. \quad (1.5.2)$$

This is the *Taylor series of f at the point c.*

In the special case where $c = 0$, we get the so-called *MacLaurin series:*

$$f(x) = f(0) + f'(0)x + \frac{1}{2!}f''(0)x^2 + \frac{1}{3!}f'''(0)x^3 + \cdots = \sum_{k=0}^{\infty} \frac{1}{k!}f^{(k)}(0)x^k. \quad (1.5.3)$$

You may be familiar with the following examples:

$$e^x = \sum_{k=0}^{\infty} \frac{x^k}{k!} = 1 + x + \frac{x^2}{2!} + \frac{x^3}{3!} + \cdots, \qquad |x| < \infty$$

$$\sin x = \sum_{k=0}^{\infty} (-1)^k \frac{x^{2k+1}}{(2k+1)!} = x - \frac{x^3}{3!} + \frac{x^5}{5!} - \frac{x^7}{7!} + \cdots, \qquad |x| < \infty$$

$$\cos x = \sum_{k=0}^{\infty} (-1)^k \frac{x^{2k}}{(2k)!} = 1 - \frac{x^2}{2!} + \frac{x^4}{4!} - \frac{x^6}{6!} + \cdots, \qquad |x| < \infty$$

$$\frac{1}{1-x} = \sum_{k=0}^{\infty} x^k = 1 + x + x^2 + x^3 + x^4 + \cdots, \qquad |x| < 1.$$

Since the computer performs only arithmetic operations, these series are actually how a computer calculates many "fancy" functions!

For example, the exponential function is calculated as

$$e^x \approx S_N \doteq \sum_{k=0}^{N} \frac{x^k}{k!}$$

for some integer N, choosing N large so that the error is sufficiently small. The value S_N is called the *partial sum* of the Taylor series.

Note that the above formula is rather "expensive" in computing time since it involves many arithmetic operations. Comparing to a single arithmetic operation, evaluating a function requires a lot more computing time. This is why in many algorithms we try to perform as few function evaluations as possible!

We now look at such an example.

Example 1.7. Compute the number e with 6 digit accuracy.

Answer. We have

$$e = e^1 = 1 + 1 + \frac{1}{2!} + \frac{1}{3!} + \frac{1}{4!} + \frac{1}{5!} + \cdots$$

And

$$\frac{1}{2!} = 0.5,$$

$$\frac{1}{3!} = 0.166667,$$

$$\frac{1}{4!} = 0.041667,$$

$$\cdots$$

$$\frac{1}{9!} = 0.0000027,$$

$$\frac{1}{10!} = 0.00000027.$$

We see that we can stop here since the term $\frac{1}{10!}$ has 6 zeros after decimal point. We now have

$$e \approx 1 + 1 + \frac{1}{2!} + \frac{1}{3!} + \frac{1}{4!} + \frac{1}{5!} + \cdots \frac{1}{9!} = 2.71828$$

Error and convergence of partial sum of Taylor series: Assume $f^{(k)}(x)$ $(0 \leq k \leq n)$ are continuous functions. Call

$$f_n(x) = \sum_{k=0}^{n} \frac{1}{k!} f^{(k)}(c)(x - c)^k$$

the partial sum of the first $n + 1$ terms in the Taylor series. We shall use $f_n(x)$ to approximate the Taylor series, with n sufficiently large.

The difference $f(x) - f_n(x)$ between the exact value of a function and its Taylor approximation is estimated by the famous *Taylor Theorem*:

$$E_{n+1} = f(x) - f_n(x) = \sum_{k=n+1}^{\infty} \frac{1}{k!} f^{(k)}(c)(x - c)^k = \frac{1}{(n+1)!} f^{(n+1)}(\xi)(x - c)^{n+1}$$

$$(1.5.4)$$

where ξ is some value between x and c. This indicates, even though the error E_{n+1} is the sum of infinitely many terms, its size is comparable to the first term in the series for E_{n+1}.

We make the following observation: A Taylor series converges rapidly if x is near c, and slowly (or not at all) if x is far away from c.

Example 1.8. (Geometric meaning for a special case of Taylor's theorem). If $n = 0$, the error estimate (1.5.4) is the same as the "Mean-Value Theorem", illustrated in Figure 1.3. Indeed, if f is smooth on the interval (a, b), then using the estimate (1.5.4) with $c = a, x = b$, we obtain

$$f(a) - f(b) = (b - a)f'(\xi), \qquad \text{for some } \xi \text{ in } (a, b).$$

This implies

$$f'(\xi) = \frac{f(b) - f(a)}{b - a}$$

If a, b are very close to each other, this formula can be used to compute an approximation for f'.

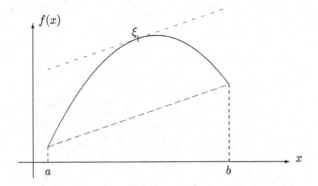

Fig. 1.3 Mean Value Theorem

1.6 Numerical Differentiations and Finite Differences

We now introduce the *finite difference approximations* for derivatives. Given $h > 0$ sufficiently small, we have 3 ways of approximating $f'(x)$:

(1) $f'(x) \approx \dfrac{f(x+h) - f(x)}{h}$ (Forward Euler)

(2) $f'(x) \approx \dfrac{f(x) - f(x-h)}{h}$ (Backward Euler)

(3) $f'(x) \approx \dfrac{f(x+h) - f(x-h)}{2h}$ (Central Finite Difference)

See Figure 1.4 for an illustration. We observe that forward Euler method takes information from the right side of x, the backward Euler takes information from the left side, while central finite difference take information from both sides.

For the second derivative f'', we can use a central finite difference approximation:

$$f''(x) \approx \frac{1}{h^2}\left(f(x+h) - 2f(x) + f(x-h)\right). \qquad (1.6.5)$$

The formula (1.6.5) is motivated as follows:

$$
\begin{aligned}
\frac{f(x+h) - 2f(x) + f(x-h)}{h^2} &= \frac{\frac{f(x+h)-f(x)}{h} - \frac{f(x)-f(x-h)}{h}}{h} \\
&\approx \frac{f'(x+h/2) - f'(x-h/2)}{h} \\
&\approx f''(x).
\end{aligned}
$$

We now study the errors in these approximations. We start with another way of writing the Taylor Series. In (1.5.1), replacing c with x, x with $x+h$, and $(x-c)$

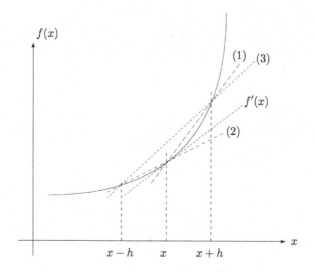

Fig. 1.4   Finite differences to approximate derivatives

with h, we get

$$f(x + h) = \sum_{k=0}^{\infty} \frac{1}{k!} f^{(k)}(x) h^k = \sum_{k=0}^{n} \frac{1}{k!} f^{(k)}(x) h^k + E_{n+1} \qquad (1.6.6)$$

where the error term is

$$E_{n+1} = \sum_{k=n+1}^{\infty} \frac{1}{k!} f^{(k)}(x) h^k = \frac{1}{(n+1)!} f^{(n+1)}(\xi) h^{n+1} \qquad (1.6.7)$$

for some ξ that lies between x and $x + h$.

This form of the Taylor series will provide the basic tool to derive error estimates, in the remainder of this course!

Local Truncation Error.   We have

$$f(x + h) = f(x) + hf'(x) + \frac{1}{2} h^2 f''(x) + \frac{1}{6} h^3 f'''(x) + \mathcal{O}(h^4),$$

$$f(x - h) = f(x) - hf'(x) + \frac{1}{2} h^2 f''(x) - \frac{1}{6} h^3 f'''(x) + \mathcal{O}(h^4).$$

Here the notation $\mathcal{O}(h^4)$ indicates a term whose size (in absolute value) is bounded by Ch^4, for some constant C. A simple computation shows

$$\frac{f(x + h) - f(x)}{h} = f'(x) + \frac{1}{2} hf''(x) + \mathcal{O}(h^2)$$

$$= f'(x) + \mathcal{O}(h^1), \qquad (1^{st} \text{ order})$$

We call this approximation *first order*, because the error is bounded by $\mathcal{O}(h^1)$. Similar computations give

$$\frac{f(x) - f(x - h)}{h} = f'(x) - \frac{1}{2} hf''(x) + \mathcal{O}(h^2)$$

$$= f'(x) + \mathcal{O}(h^1), \qquad (1^{st} \text{ order})$$

and

$$\frac{f(x+h) - f(x-h)}{2h} = f'(x) - \frac{1}{6}h^2 f'''(x) + \mathcal{O}(h^2)$$
$$= f'(x) + \mathcal{O}(h^2), \quad (2^{nd} \text{ order})$$

and finally

$$\frac{f(x+h) - 2f(x) + f(x-h)}{h^2} = f''(x) + \frac{1}{12}h^2 f^{(4)}(x) + \mathcal{O}(h^4)$$
$$= f''(x) + \mathcal{O}(h^2), \quad (2^{nd} \text{ order}).$$

In conclusion, one-sided Euler approximations are first order, while central finite difference approximations are second order.

1.7 Homework Problems for Chapter 1

Problem 1

(a). Convert the binary numbers to decimal numbers.

i). $(110111001.101011101)_2$ ii). $(1001100101.01101)_2$

(b). Convert the decimal numbers to binary. Keep 10 fractional points.

i). $(100.01)_{10}$ (ii). $(64.625)_{10}$

(c). Write $(64.625)_{10}$ into normalized scientific notation in binary. You could use the result in the previous problem. Then determine how it would look like in a 32-bit computer, using a single-precision floating point representation.

Problem 2: Error Propagation

Perform a detailed study for the error propagation in the computation $z = xy$. Let $fl(x) = x(1 + \delta_x)$ and $fl(y) = y(1 + \delta_y)$ where $fl(x)$ is the floating point representation of x. Find the expression for the absolute error and the relative error in the answer $fl(z)$.

Problem 3: Loss of Significance

(a). For some values of x, the function $f(x) = \sqrt{x^2 + 1} - x$ cannot be computed accurately in a computer by using this formula. Explain why and demonstrate it with an example. Then find a way around the difficulty.

(b). Explain why the function

$$f(x) = \frac{1}{\sqrt{x + 2} - \sqrt{x}}$$

cannot be computed accurately in a computer when x is large (using the above formula). Find a way around the problem.

Problem 4

(a). Derive the following Taylor series for $(1 + x)^n$ (this is also known as the *binomial series*):

$$(1 + x)^n = 1 + nx + \frac{n(n - 1)}{2!}x^2 + \frac{n(n - 1)(n - 2)}{3!}x^3 + \cdots \quad (x^2 < 1)$$

Write out its particular form when $n = 2$, $n = 3$, and $n = 1/2$. Then use the last form to compute $\sqrt{1.0001}$ correct to 15 decimal places (rounded).

(b). Use the answer in part (a) to obtain a series for $(1 + x^2)^{-1}$.

Problem 5: Matlab Exercises

The goal of this exercise is to get started with Matlab. You will go through:

- Matrices, vectors, solutions of systems of linear equations;
- simple plots;
- Use of Matlab's own help functions.

To get started, you can read the first two chapters in "*A Practical Introduction to Matlab*" by Gockenbach. You can find the introduction at the web-page:

http://www.math.mtu.edu/~msgocken/intro/intro.pdf

Read through the Chapter "Introduction" and "Simple calculations and graphs".

How to do it:

Find a computer with Matlab installed in it. Start Matlab, by either clicking on the appropriate icon (Windows or Mac), or by typing in matlab (unix or linux). You should get a command window on the screen.

Go through the examples in Gockenbach's notes.

Help!

As we will see later, Matlab has many build-in numerical functions. One of them is a function that does polynomial interpolation. But what is the name, and how to use it? You can use lookfor to find the function, and help to get a description on how to use it.

lookfor keyword: Look after the "keyword" among Matlab functions.

There is nothing to turn in here. Just have fun and get used to Matlab!

Problem 6: A Study on Loss of Significance

In this problem we shall consider the effect of computer arithmetic in the evaluation of the quadratic formula

$$r_{1,2} = \frac{-b \pm \sqrt{b^2 - 4ac}}{2a} \qquad (1.7.8)$$

for the roots of $p(x) = ax^2 + bx + c$.

(a). Write a Matlab function, call it quadroots, that takes a, b, c as input and returns the two roots as output. The function may start like this:

```
function [r1, r2] = quadroots(a,b,c)
% input: a, b, c: coefficients for the polynomial ax^2+bx+c=0.
% output: r1, r2: The two roots for the polynomial.
```

Use the formula in (1.7.8) for the computation. Run it on a few test problems to make sure that it works before you go ahead with the following problems.

(b). Test your `quadroots` for the following polynomials:

- $2x^2 + 6x - 3$
- $x^2 - 14x + 49$
- $3x^2 - 123454321x + 2$

What do you get? Why are the results pretty bad for the last polynomial? Can you explain?

(c). The product of the roots of $ax^2 + bx + c$ is of course c/a. Use this fact to improve your program, call it `smartquadroots`, to see if you get better results.

Matlab has a command `roots(p)` which returns the roots of the polynomial with coefficients in **p**. `roots` is smarter than `quadroots` and will give accurate answers here. You can use it to check your results from `smartquadroots` function.

What to hand in: Hand in two Matlab function m-files `quadroots.m` and `smartquadroots.m`, and the results you get, together with your comments.

Chapter 2

Polynomial Interpolation

2.1 Introduction

In this chapter we study how to interpolate a data set with a polynomial. By a data set, we mean a set of points in the (x, y) plane.

Problem description: Consider $(n + 1)$ points in the plane, say (x_i, y_i), with $i = 0, 1, 2, \cdots, n$, and assume that the x_i are all distinct:

$$x_0 < x_1 < x_2 < \cdots < x_n .$$

We want to find a polynomial of degree n, which can be written in general form as

$$P_n(x) = a_n x^n + a_{n-1} x^{n-1} + \cdots + a_1 x + a_0,$$

which interpolates all points in the data set, namely

$$P_n(x_i) = y_i, \qquad i = 0, 1, 2, \cdots, n.$$

Of course, the polynomial P_n will be determined as soon as we find the coefficients $a_n, a_{n-1}, \cdots, a_1, a_0$.

Notice that the total number of data points is equal to the number of coefficients a_i which we need to determine. This is one more than the degree of the polynomial.

Why are we interested in doing this? Here are some possible reasons:

- A function may be known only on a discrete data set, as a result of some lab experiments. We wish to make a reasonable guess for the values of this function at all other points.
- From basic calculus, it is easier to compute integrals and derivatives of a polynomial, rather than a general function. Because of this, we may want to approximate a (possibly very complicated) function by a polynomial.

We start with a simple example.

Example 2.1. Given the data set in the table

x_i	0	1	2/3
y_i	1	0	0.5

,

interpolate these data with a polynomial of degree 2.

Note that all these data satisfy the relation $y = \cos(\pi x/2)$, i.e.,

$$y_i = \cos(\pi x_i/2).$$

Answer. Let

$$P_2(x) = a_2 x^2 + a_1 x + a_0.$$

We need to find the coefficients a_2, a_1, a_0. We shall treat these three coefficients as the unknowns, and write out all the equations they must satisfy. From the data set we obtain:

$$x = 0, \; y = 1 : \quad P_2(0) = a_0 = 1,$$
$$x = 1, \; y = 0 : \quad P_2(1) = a_2 + a_1 + a_0 = 0,$$
$$x = 2/3, \; y = 0.5 : \quad P_2(2/3) = (4/9)a_2 + (2/3)a_1 + a_0 = 0.5.$$

We thus have 3 equations in 3 unknowns, and we thus expect to find a unique solution. The above system of equations can be written in matrix-vector form

$$\begin{pmatrix} 0 & 0 & 1 \\ 1 & 1 & 1 \\ \frac{4}{9} & \frac{2}{3} & 1 \end{pmatrix} \begin{pmatrix} a_2 \\ a_1 \\ a_0 \end{pmatrix} = \begin{pmatrix} 1 \\ 0 \\ 0.5 \end{pmatrix}.$$

This method is originally due to Vandermonde, and the coefficient matrix is thus called the *Vandermonde matrix*. You can solve this system using Matlab (see sample codes below, and also the homework problem), or do it by hand with Gaussian elimination. The solution is

$$a_2 = -\frac{3}{4}, \quad a_1 = -\frac{1}{4}, \quad a_0 = 1.$$

Hence the interpolating polynomial is

$$P_2(x) = -\frac{3}{4}x^2 - \frac{1}{4}x + 1.$$

Sample Matlab Code. This example can be easily coded in Matlab.

```
>> X = [1,0,0;1,1,1;1,2/3,4/9]
% van der Monde matrix
X =
    1.0000         0         0
    1.0000    1.0000    1.0000
    1.0000    0.6667    0.4444

>> y = [1;0;1/2]

y =
```

```
   1.0000
        0
   0.5000
```

```
>> a = X\y % This computes the value a, as the solution.
% One can also use the command:   a=inv(X)*y
% which computes the same thing,
% but with a different approach.
```

```
a =
    1.0000
   -0.2500
   -0.7500
```

The coefficients are now stored in the variable a.

To plot the interpolating points, one can do

```
>> x = [0;1;2/3];
>> y = [1;0;1/2];
>> plot(x,y,'*'); grid;
```

The plot is given in Figure 2.1. We see three data points in the plot, marked with '*'.

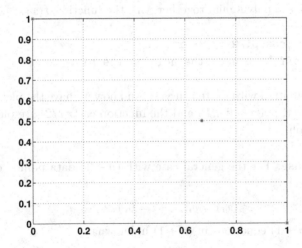

Fig. 2.1   Plot of the data set.

To plot the interpolating polynomial, together with the data set, one can do the following:

```
>> hold on % keep the previous plot
>> t = [0:0.01:1];
>> p2 = a(1)+a(2)*t+a(3)*t.^2;
>> plot(t,p2);
```

The plot is given in Figure 2.2.

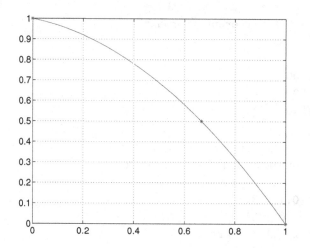

Fig. 2.2   Plot of the interpolating polynomial together with the data set.

One can plot the polynomial together with the function $f(x) = \cos(\pi x/2)$, as

```
>> plot(t,cos(pi/2*t),'--r')
>> hold off % release the hold on the plots
```

This function is plotted with dotted curves. And now we have the plot in Figure 2.3, where the data set, polynomial P_2 and the function $\cos(\pi x/2)$ are plotted together in the same graph.

The general case. For the general case with $(n+1)$ data points, one has $n+1$ requirements:

$$P_n(x_i) = y_i, \qquad i = 0, 1, 2, \cdots, n.$$

We thus have $(n+1)$ equations in $(n+1)$ unknowns:

$$P_n(x_0) = y_0 : x_0^n a_n + x_0^{n-1} a_{n-1} + \cdots + x_0 a_1 + a_0 = y_0,$$
$$P_n(x_1) = y_1 : x_1^n a_n + x_1^{n-1} a_{n-1} + \cdots + x_1 a_1 + a_0 = y_1,$$
$$\vdots$$
$$P_n(x_n) = y_n : x_n^n a_n + x_n^{n-1} a_{n-1} + \cdots + x_n a_1 + a_0 = y_n.$$

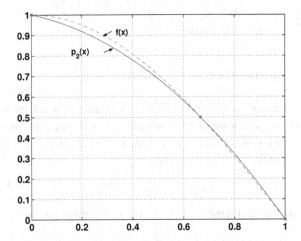

Fig. 2.3 Plot of the interpolating polynomial together with the data set and the function that generated the data.

Writing this system of equations in matrix-vector form, we obtain

$$\begin{pmatrix} x_0^n & x_0^{n-1} & \cdots & x_0 & 1 \\ x_1^n & x_1^{n-1} & \cdots & x_1 & 1 \\ \vdots & \vdots & \ddots & \vdots & \vdots \\ x_n^n & x_n^{n-1} & \cdots & x_n & 1 \end{pmatrix} \begin{pmatrix} a_n \\ a_{n-1} \\ \vdots \\ a_0 \end{pmatrix} = \begin{pmatrix} y_0 \\ y_1 \\ \vdots \\ y_n \end{pmatrix},$$

or with compact notation

$$\mathbf{X}\,\vec{a} = \vec{y},$$

where

$\mathbf{X}$: $(n+1) \times (n+1)$ is the Vandermonde matrix, obtained from the given data points x_i,

$\vec{a}$: unknown vector, with length $(n+1)$,

$\vec{y}$: given vector, with length $(n+1)$.

A well-known theorem states that, if all the x_i's are distinct, then the matrix $\mathbf{X}$ is invertible, therefore the solution vector $\vec{a}$ is uniquely determined.

In Matlab, the Vandermonde matrix can be generated using the command `vander(x)`, where x is the vector containing the interpolation points x=$[x_0, x_1, x_2, \cdots, x_n]$.

There are some bad news. It is known that $\mathbf{X}$ has very large condition number for large values of n. Therefore, the system is hard to solve when n is large. (Condition number and its significance will be discussed in detail later in Chapter 6.) This fact alone renders this method of little use, for n large.

Other more efficient and elegant methods, which we will study in the following sections, include

- Lagrange polynomials,
- Newton's divided differences.

2.2   Lagrange Interpolation Polynomials

The Lagrange interpolating polynomials are constructed starting with a basis of polynomial functions, which are locally supported in the discrete sense.

More precisely, let $n+1$ distinct points $x_0, x_1, \cdots, x_n$ be given. We then define a family of corresponding *cardinal functions*

$$l_0, l_1, \cdots, l_n \in \mathcal{P}^n,$$

which are polynomials of degree n, and satisfy the properties

$$l_i(x_j) = \delta_{ij} = \begin{cases} 1 \,, i = j \\ 0 \,, i \neq j \end{cases} \qquad i = 0, 1, \cdots, n. \qquad (2.2.1)$$

Here δ_{ij} is called the *Kronecker's delta*.

In other words, the cardinal function $l_i(x)$ takes the value 1 for $x = x_i$, but it takes the value 0 at all other interpolating points $x = x_j$ with $j \neq i$. We call the property (2.2.1) "locally supported in a discrete sense".

The cardinal functions $l_i(x)$ can be written out explicitly, using this elegant formula:

$$l_i(x) = \prod_{j=0, j \neq i}^{n} \left(\frac{x - x_j}{x_i - x_j} \right)$$

$$= \frac{x - x_0}{x_i - x_0} \cdot \frac{x - x_1}{x_i - x_1} \cdots \frac{x - x_{i-1}}{x_i - x_{i-1}} \cdot \frac{x - x_{i+1}}{x_i - x_{i+1}} \cdots \frac{x - x_n}{x_i - x_n}. \qquad (2.2.2)$$

Indeed, this formula guarantees that all conditions in (2.2.1) are satisfied.

Next, the *Lagrange form of the interpolation polynomial* is

$$P_n(x) = \sum_{i=0}^{n} l_i(x) \cdot y_i. \qquad (2.2.3)$$

Using (2.2.1), it is easy to check the interpolation property:

$$P_n(x_j) = \sum_{i=0}^{n} l_i(x_j) \cdot y_i = l_j(x_j) y_j = y_j, \qquad \text{for all } j = 0, 1, \ldots, n.$$

Note that the cardinal functions $l_i(x)$ are determined as soon as the points $(x_i)_{i=0}^{n}$ are given (i.e., they do not depend on the values y_i). The values $(y_i)_{i=0}^{n}$ are only used when we construct the polynomial in (2.2.3).

Example 2.2. Consider again (same as in Example 2.1)

$$\begin{array}{c|c|c|c} x_i & 0 & 2/3 & 1 \\ \hline y_i & 1 & 0.5 & 0 \end{array}.$$

Write out the Lagrange form of the interpolating polynomial.

Answer. The data set corresponds to

$$x_0 = 0, \quad x_1 = 2/3, \quad x_2 = 1,$$

and

$$y_0 = 1, \quad y_1 = 0.5, \quad y_2 = 0.$$

We first compute the cardinal functions. Using x_0, x_1, x_2 we obtain

$$l_0(x) = \frac{(x - x_1)(x - x_2)}{(x_0 - x_1)(x_0 - x_2)} = \frac{(x - 2/3)\,(x - 1)}{(0 - 2/3)\,(0 - 1)} = \frac{3}{2}\left(x - \frac{2}{3}\right)(x - 1)$$

$$l_1(x) = \frac{(x - x_0)(x - x_2)}{(x_1 - x_0)(x_1 - x_2)} = \frac{(x - 0)\,(x - 1)}{(2/3 - 0)\,(2/3 - 1)} = -\frac{9}{2}x(x - 1)$$

$$l_2(x) = \frac{(x - x_0)(x - x_1)}{(x_2 - x_0)(x_2 - x_1)} = \frac{(x - 0)\,(x - 2/3)}{(1 - 0)\,(1 - 2/3)} = 3x\left(x - \frac{2}{3}\right).$$

The Lagrange polynomial is now defined as

$$\begin{aligned} P_2(x) &= l_0(x)y_0 + l_1(x)y_1 + l_2(x)y_2 \\ &= \frac{3}{2}\left(x - \frac{2}{3}\right)(x - 1) - \frac{9}{2}x(x - 1)(0.5) + l_2(x) \cdot 0 \\ &= -\frac{3}{4}x^2 - \frac{1}{4}x + 1. \end{aligned}$$

Notice that, after some simplifications, we get the same answer as in Example 2.1. Do you think this is a coincidence or it should always happen?

Pros and cons of Lagrange polynomial interpolation:

(+) Elegant formula.

(+) Easily adjustable if the values y_i are changed (but the x_i remain the same).

(−) Slow to compute, since each $l_i(x)$ requires a different calculation.

(−) Not flexible: if one changes a point x_j, or inserts an additional point x_{n+1} (maybe to improve accuracy), one must compute all polynomials l_i over again.

In the next section we study a method that allows the flexibility of adding more points to the data set.

2.3 Newton's Divided Differences

Consider a data set

x_i	x_0	x_1	$\cdots$	x_n
y_i	y_0	y_1	$\cdots$	y_n

We shall describe this interpolation algorithm, which has a recursive form.

Main idea: Given a polynomial $P_{k-1}(x)$ of degree $k-1$ that interpolates k data points, one computes the next polynomial $P_k(x)$ that interpolates one extra point, i.e., $k+1$ data points. The polynomial P_k has degree $k+1$, and is obtained from P_{k-1} by adding an extra term. We start with P_0, then proceed by induction.

- For $n = 0$, we define $P_0(x) = a_0 = y_0$.
- For $n = 1$, we set

$$P_1(x) \;=\; P_0(x) + a_1(x - x_0) \;=\; y_0 + a_1(x - x_0), \qquad (2.3.4)$$

where a_1 must be determined. Note that this particular form (2.3.4) automatically guarantees the interpolating property $P_1(x_0) = y_0$, because the additional term $a_1(x - x_0)$ vanishes at $x = x_0$ for any choice of a_1.

Requiring that P_1 interpolates the new point x_1, i.e., $y_1 = P_1(x_1)$, we obtain

$$y_1 \;=\; a_0 + a_1(x_1 - x_0).$$

This gives us a formula for the coefficient a_1, namely

$$a_1 = \frac{y_1 - y_0}{x_1 - x_0}. \qquad (2.3.5)$$

- For $n = 2$, we set

$$P_2(x) = P_1(x) + a_2(x - x_0)(x - x_1). \qquad (2.3.6)$$

Note that the additional term $a_2(x - x_0)(x - x_1)$ vanishes at both $x = x_0$ and $x = x_1$. Therefore the interpolation properties

$$P_2(x_0) = P_1(x_0) = P_0(x_0) = y_0, \qquad P_2(x_1) = P_1(x_1) = y_1$$

will certainly hold for any choice of a_2.

Again, the coefficient a_2 is determined by imposing the interpolation property at the new point $x = x_2$, namely: $y_2 = P_2(x_2)$. This yields

$$y_2 \;=\; P_1(x_2) + a_2(x_2 - x_0)(x_2 - x_1),$$

which leads to the formula

$$a_2 \;=\; \frac{y_2 - P_1(x_2)}{(x_2 - x_0)(x_2 - x_1)}.$$

We would like to express a_2 in a different way, and obtain a recursive formula valid for the general case. Using (2.3.4) and (2.3.5), we get

$$P_1(x_2) = y_0 + \frac{y_1 - y_0}{x_1 - x_0}(x_2 - x_0)$$

$$= y_0 + \frac{y_1 - y_0}{x_1 - x_0}(x_2 - x_1) + \frac{y_1 - y_0}{x_1 - x_0}(x_1 - x_0)$$

$$= y_1 + \frac{y_1 - y_0}{x_1 - x_0}(x_2 - x_1).$$

Then, a_2 can be rewritten as

$$a_2 = \frac{y_2 - y_1 - \frac{y_1 - y_0}{x_1 - x_0}(x_2 - x_1)}{(x_2 - x_0)(x_2 - x_1)} = \frac{\frac{y_2 - y_1}{x_2 - x_1} - \frac{y_1 - y_0}{x_1 - x_0}}{x_2 - x_0}. \tag{2.3.7}$$

General expression for a_k: Assume that $P_{k-1}(x)$ interpolates (x_i, y_i) for $i = 0, 1, \cdots, k-1$. We will find $P_k(x)$ that interpolates (x_i, y_i) for $i = 0, 1, \cdots, k$, i.e., P_k interpolates also the additional point (x_k, y_k). The new polynomial $P_k(x)$ takes the form

$$P_k(x) = P_{k-1}(x) + a_k(x - x_0)(x - x_1) \cdots (x - x_{k-1}), \tag{2.3.8}$$

where

$$a_k = \frac{y_k - P_{k-1}(x_k)}{(x_k - x_0)(x_k - x_1) \cdots (x_k - x_{n-1})}. \tag{2.3.9}$$

One can easily check that such polynomial takes the correct value y_i at each point x_i. Indeed, because of its particular form, the new term $a_k(x - x_0)(x - x_1) \cdots (x - x_{k-1})$, vanishes at all the points x_i for $i = 0, 1, \cdots, k-1$. Therefore, for $i = 0, 1, \ldots, k-1$ we have $P_k(x_i) = P_{k-1}(x_i) = y_i$. Finally, at the new point x_n we have $P_k(x_k) = y_k$, since a_k was chosen as in (2.3.9), which guarantees this interpolation property.

One can conclude by induction that the polynomial P_n interpolates all $(n+1)$ data points. This now gives the *Newton's form of the interpolating polynomial*:

$$P_n(x) = a_0 + a_1(x - x_0) + a_2(x - x_0)(x - x_1) + \cdots$$
$$+ a_n(x - x_0)(x - x_1) \cdots (x - x_{n-1}),$$

where the coefficients a_k are determined by (2.3.9).

There is a more elegant way of computing these coefficients a_k, which we now describe. We introduce some quantities called *Newton's divided differences*, which will be defined recursively. The recursion is initiated with

$$f[x_i] = y_i, \qquad i = 0, 1, 2, \cdots.$$

Then

$$f[x_0, x_1] = \frac{f[x_1] - f[x_0]}{x_1 - x_0},$$

$$f[x_1, x_2] = \frac{f[x_2] - f[x_1]}{x_2 - x_1},$$

$$f[x_0, x_1, x_2] = \frac{f[x_1, x_2] - f[x_0, x_1]}{x_2 - x_0},$$

$$f[x_1, x_2, x_3] = \frac{f[x_2, x_3] - f[x_1, x_2]}{x_3 - x_1}.$$

By now, the pattern should be clear. For the general step, we have

$$f[x_0, x_1, \cdots, x_k] = \frac{f[x_1, x_2, \cdots, x_k] - f[x_0, x_1, \cdots, x_{k-1}]}{x_k - x_0}. \qquad (2.3.10)$$

For $k = 0, 1, \ldots, n$, the constants a_k in Newton's polynomial are computed by

$$a_0 = f[x_0],$$

$$a_1 = f[x_0, x_1],$$

$$\cdots$$

$$a_n = f[x_0, x_1, \cdots, x_n].$$

Proof of (2.3.10): (optional) The proof is achieved by induction. The formula is clearly true for $n = 0, 1, 2$. Now, assume that it holds for $n = k - 1$, i.e., one can use it to write the polynomial of degree $k - 1$, to interpolate k points, in Newton's form. We show that it also holds for $n = k$. By induction, the formula will thus be true for all n.

Let $P_{k-1}(x)$ interpolate the points $(x_i, y_i)_{i=0}^{k-1}$, and let $q(x)$ interpolate the points $(x_i, y_i)_{i=1}^{k}$. Note that, compared to P_{k-1}, the function $q(x)$ does not interpolate (x_0, y_0). Instead it interpolates the extra point (x_k, y_k). Both P_{k-1} and $q(x)$ are polynomials of degree $k - 1$. By our inductive assumption, formula (2.3.10) holds in Newton's form, i.e.,

$$P_{k-1}(x) = P_{k-2}(x) + f[x_0, \cdots, x_{k-1}](x - x_0)(x - x_1) \cdots (x - x_{k-2})$$

$$= f[x_0, \cdots, x_{k-1}]x^{k-1} + \text{(l.o.t.)} \ \text{(i.e. lower order terms)} \qquad (2.3.11)$$

and

$$q(x) = f[x_1, \cdots, x_k]x^{k-1} + \text{(l.o.t.)}. \qquad (2.3.12)$$

We now set

$$P_k = q(x) + \frac{x - x_k}{x_k - x_0}\left(q(x) - P_{k-1}(x)\right).$$

We claim that $P_k(x)$ interpolates all the points $(x_i, y_i)_{i=0}^k$.

To check this claim, we go through all the points x_i with $i = 0, 1, 2, \cdots, k$, as

$$i = 1, 2, \cdots, k : \quad q(x_i) = P_{k-1}(x_i) = y_i, \quad P_k(x_i) = y_i$$

$$i = 0 : \quad P_k(x_0) = q(x_0) + \frac{x_0 - x_k}{x_k - x_0}(q(x_0) - y_0) = y_0,$$

$$i = k : \quad P_k(x_k) = q(x_k) + 0 = y_k.$$

By using (2.3.11)-(2.3.12), we can now write

$$P_k(x) = f[x_1, \cdots, x_k]x^{k-1} + \text{l.o.t.}$$

$$+ \frac{x - x_k}{x_k - x_0}\left[f[x_0, \cdots, x_{k-1}]x^{k-1} + (\text{l.o.t.})\right]$$

$$= \frac{f[x_1, \cdots, x_k] - f[x_0, \cdots, x_{k-1}]}{x_k - x_0} \cdot x^k + (\text{l.o.t.}).$$

We compare this with the Newton's expression for P_k

$$P_k(x) = P_{k-1}(x) + f[x_0, \cdots, x_k](x - x_0)\cdots(x - x_{k-1})$$

$$= f[x_0, \cdots, x_k]x^k + (\text{l.o.t.}).$$

Since these are the same polynomial (the uniqueness of interpolating polynomials will be proved later), they must have the same coefficient for the leading term x^k, i.e.,

$$f[x_0, \cdots, x_k] = \frac{f[x_1, \cdots, x_k] - f[x_0, \cdots, x_{k-1}]}{x_k - x_0},$$

proving (2.3.10). □

Computation of the divided differences: We compute the $f[\cdots]$'s through the following table:

x_0	$f[x_0] = y_0$				
x_1	$f[x_1] = y_1$	$f[x_0, x_1]$ $= \frac{f[x_1]-f[x_0]}{x_1-x_0}$			
x_2	$f[x_2] = y_2$	$f[x_1, x_2]$ $= \frac{f[x_2]-f[x_1]}{x_2-x_1}$	$f[x_0, x_1, x_2]$		
$\vdots$	$\vdots$	$\vdots$	$\vdots$	$\ddots$	
x_n	$f[x_n] = y_n$	$f[x_{n-1}, x_n]$ $= \frac{f[x_n]-f[x_{n-1}]}{x_n-x_{n-1}}$	$f[x_{n-2}, x_{n-1}, x_n]$	$\cdots$	$f[x_0, x_1, \cdots, x_n]$

Note that this yields a triangular table of computed data, due to the fact that each divided difference is computed using two divided differences from the previous level. The algorithm should be obvious from the tabular setting.

Example 2.3. Use Newton's divided difference to write the polynomial that interpolates the data

x_i	0	1	2/3	1/3
y_i	1	0	1/2	0.866

Answer. We set up the triangular table for the computation:

$$
\begin{array}{c||c}
0 & \boxed{1} \\
1 & 0 & \frac{0-1}{1-0} = \boxed{-1} \\
2/3 & 0.5 & \frac{0.5-0}{2/3-1} = -1.5 & \frac{-1.5-(-1)}{2/3-0} = \boxed{-0.75} \\
1/3 & 0.866 & \frac{0.866-0.5}{1/3-2/3} = -1.098 & \frac{-1.098-(-1.5)}{1/3-1} = -0.603 & \frac{-0.603-(-0.75)}{1/3-0} = \boxed{0.441}
\end{array}
$$

So we have

$$a_0 = 1, \quad a_1 = -1, \quad a_2 = -0.75, \quad a_3 = 0.441,$$

and

$$x_0 = 0, \quad x_1 = 1, \quad x_2 = 2/3, \quad x_3 = 1/3.$$

The interpolating polynomial in Newton's form is thus

$$P_3(x) = a_0 + a_1(x - x_0) + a_2(x - x_0)(x - x_1) + a_3(x - x_0)(x - x_1)(x - x_2)$$
$$= \boxed{1} + \boxed{-1}\, x + \boxed{-0.75}\, x(x-1) + \boxed{0.441}\, x(x-1)(x-2/3).$$

Flexibility of Newton's form: When Newton's form is used, it is easy to add more data points to interpolate. Moreover, all the previous computations can still be used.

To see this, assume that in the previous example we want to interpolate the additional data point $(x_4, y_4) = (0.5, 1)$. We can keep all the earlier work, and add one more line in the table, to get a_4. Try it by yourself!

Nested form: Newton's polynomial can also be written as

$$P_n(x) = a_0 + a_1(x - x_0) + a_2(x - x_0)(x - x_1) + \cdots$$
$$+ a_n(x - x_0)(x - x_1) \cdots (x - x_{n-1})$$
$$= a_0 + (x - x_0)\,(a_1 + (x - x_1)(a_2 + (x - x_2)(a_3 + \cdots + a_n(x - x_{n-1}))))$$

This is called the *nested form of Newton's polynomial.*

This form is very effective to evaluate, in a computer program. Given the data x_i and a_i for $i = 0, 1, \cdots, n$, one can evaluate the Newton's polynomial $p = P_n(x)$ by the following algorithm (pseudo-code):

$$p = a_n$$
$$\text{for } k = n - 1, n - 2, \cdots, 0$$
$$\qquad p = p(x - x_k) + a_k$$
$$\text{end}$$

This requires only $3n$ flops of computation, which is very efficient.

Before stating a theorem on the existence and uniqueness for polynomial interpolation, we recall the Fundamental Theorem of Algebra.

Theorem 2.1. *(Fundamental Theorem of Algebra) Every polynomial $p(x)$ of degree n, which is not identically zero, has exactly n roots (counting multiplicities). These roots may be real or complex. In particular, if a polynomial of degree n has more than n distinct roots, then it must be identically zero.*

Using Theorem 2.1, we can prove:

Theorem 2.2. *(Existence and Uniqueness of Polynomial Interpolation) Consider data points $(x_i, y_i)_{i=0}^n$, with the x_i's all distinct. Then there exists one and only polynomial $P_n(x)$ of degree $\leq n$ such that*

$$P_n(x_i) = y_i, \qquad i = 0, 1, \cdots, n.$$

Proof. The existence of such a polynomial is obvious, since we can construct it using any one of the methods previously described.

Regarding uniqueness, we prove it by contradiction. Assume that we have two different polynomials, call them $p(x)$ and $q(x)$, of degree $\leq n$, both interpolate the data, i.e.,

$$p(x_i) = y_i, \qquad q(x_i) = y_i, \qquad i = 0, 1, \cdots, n.$$

Now let $g(x) = p(x) - q(x)$, which will be a polynomial of degree $\leq n$. Furthermore, we have

$$g(x_i) = p(x_i) - q(x_i) = y_i - y_i = 0, \qquad i = 0, 1, \cdots, n.$$

So $g(x)$ has $n + 1$ zeros. By the Fundamental Theorem of Algebra, we must have $g(x) \equiv 0$, therefore $p(x) \equiv q(x)$. This implies uniqueness. $\qquad\square$

2.4 Errors in Polynomial Interpolation

Consider a function $f(x)$ on the interval $a \leq x \leq b$, and a set of distinct points $x_i \in [a, b]$, $i = 0, 1, \cdots, n$. Let $P_n(x)$ be a polynomial of degree $\leq n$ that interpolates $f(x)$ at the points x_i, i.e.,

$$P_n(x_i) = f(x_i), \qquad i = 0, 1, \cdots, n.$$

We then define the *error function* to be the difference between the function f and the approximating polynomial P_n:

$$e(x) = f(x) - P_n(x), \qquad x \in [a, b].$$

In order to estimate the size of this error, we shall use the following theorem.

Theorem 2.3. (Interpolation Error Theorem). *For every $x \in [a, b]$ one can find some value $\xi \in [a, b]$ such that*

$$e(x) = \frac{1}{(n+1)!} f^{(n+1)}(\xi) \cdot \prod_{i=0}^{n} (x - x_i). \qquad (2.4.13)$$

Proof. If f is a polynomial of degree n, i.e., $f \in \mathcal{P}_n$, then by the Uniqueness Theorem of polynomial interpolation we must have $f(x) = P_n(x)$, hence $e(x) \equiv 0$. In this case, the derivative of order $n + 1$ is $f^{(n+1)}(\xi) = 0$ for every ξ, and the identity (2.4.13) is clear.

Now assume $f \notin \mathcal{P}_n$. If $x = x_i$ for some i, we have $e(x_i) = f(x_i) - P_n(x_i) = 0$, and the result holds.

Now consider $x \neq x_i$ for any i. We define

$$W(x) = \prod_{i=0}^{n} (x - x_i) \quad \in \mathcal{P}_{n+1},$$

then it holds

$$W(x_i) = 0, \qquad W(x) = x^{n+1} + \cdots, \qquad W^{(n+1)}(x) = (n+1)!.$$

Fix a point y such that $a \leq y \leq b$ with $y \neq x_i$ for every i. We define the constant

$$c = \frac{f(y) - P_n(y)}{W(y)},$$

and consider the function

$$\varphi(x) = f(x) - P_n(x) - cW(x).$$

We now look at all the zeros of this function φ. Every point x_i is a zero, because

$$\varphi(x_i) = f(x_i) - P_n(x_i) - cW(x_i) = 0, \qquad i = 0, 1, \cdots, n.$$

The point y is also a zero, because

$$\varphi(y) = f(y) - P_n(y) - cW(y)$$
$$= f(y) - P_n(y) - \frac{f(y) - P_n(y)}{W(y)} W(y)$$
$$= 0.$$

So, φ has at least $(n + 2)$ distinct zeros.

Here is the key idea of the proof: if a function has at least k distinct zeros, then its derivative will have at least $k - 1$ zeros. Indeed, between any two zeros of the function by Rolle's theorem there must be a zero of the derivative.

By using the previous argument several times, we obtain

$$
\begin{array}{lll}
\varphi \text{ has at least} & n + 2 & \text{zeros on } [a, b]. \\
\varphi' \text{ has at least} & n + 1 & \text{zeros on } [a, b]. \\
\varphi'' \text{ has at least} & n & \text{zeros on } [a, b]. \\
& \vdots & \\
\varphi^{(n+1)} \text{ has at least} & 1 & \text{zero on } [a, b].
\end{array}
$$

Let ξ be a point in the interval $[a, b]$ where this derivative vanishes, so that

$$\varphi^{(n+1)}(\xi) = 0.$$

Using the identity $P_n^{(n+1)}(x) \equiv 0$, we find

$$\varphi^{(n+1)}(\xi) = f^{(n+1)}(\xi) - P_n^{(n+1)}(x) - cW^{(n+1)}(\xi)$$
$$= f^{(n+1)}(\xi) - cW^{(n+1)}(\xi)$$
$$= 0.$$

Observing that

$$W^{(n+1)}(x) = (n + 1)! \qquad \text{for all} \quad x,$$

we obtain

$$f^{(n+1)}(\xi) = cW^{(n+1)}(\xi) = \frac{f(y) - P_n(y)}{W(y)}(n + 1)!.$$

Writing x in place of y, from the above identity it follows

$$e(x) = f(x) - P_n(x)$$
$$= \frac{1}{(n + 1)!} f^{(n+1)}(\xi) W(x)$$
$$= \frac{1}{(n + 1)!} f^{(n+1)}(\xi) \cdot \prod_{i=0}^{n} (x - x_i),$$

for some $\xi \in [a, b]$, depending on x. $\qquad\qquad\qquad\qquad\qquad\qquad\qquad\square$

Using the representation Theorem 2.3, we can establish error bounds for polynomial interpolation. An upper bound for the error is a quantity that is larger than the absolute value of the error.

Example 2.4. Consider an interval $[a, b]$, and choose $n = 1$, $x_0 = a, x_1 = b$. We thus interpolate a function f with a polynomial P_1 of degree 1. For any $x \in [a, b]$, to find an upper bound for the error we write

$$|e(x)| = \frac{1}{2} |f''(\xi)| \cdot |(x - a)(x - b)|$$

$$\leq \frac{1}{2} \|f''\|_\infty \frac{(b - a)^2}{4}$$

$$= \frac{1}{8} \|f''\|_\infty (b - a)^2.$$

Here we used the estimate

$$\max_{x \in [a,b]} |(x - a)(x - b)| = \left| \left(\frac{a + b}{2} - a \right) \left(\frac{a + b}{2} - b \right) \right| = \frac{(b - a)^2}{4},$$

and the notation $\|f''\|_\infty = \max_{\xi \in [a,b]} |f''(\xi)|$.

We observe that different choices of the nodes x_i would give different errors.

Uniform grid is a grid which evenly distributes the interpolating points in space. Given an interval $[a, b]$, we distribute $n + 1$ points uniformly, by setting

$$h = \frac{b - a}{n}, \qquad x_i = a + ih, \qquad i = 0, 1, \cdots, n.$$

The next Lemma yields an error bound for polynomial interpolation, when a uniform grid is used.

Lemma 2.1. *For $x \in [a, b]$, it holds*

$$\prod_{i=0}^{n} |x - x_i| \leq \frac{1}{4} h^{n+1} \cdot n!.$$

Proof. If $x = x_i$ for some i, then the product is 0, so it trivially holds. Now assume $x_i < x < x_{i+1}$ for some i. We have

$$\max_{x_i < x < x_{i+1}} |(x - x_i)(x - x_{i+1})| = \frac{1}{4}(x_{i+1} - x_i)^2 = \frac{h^2}{4}.$$

Now consider the other terms in the product, say $x - x_j$, for either $j > i + 1$ or $j < i$. Then $|x - x_j| \leq h(j - i)$ for $j > i + 1$ and $|x - x_j| \leq h(i + 1 - j)$ for $j < i$. In all cases, the product of these terms are bounded by $h^{n-1} n!$, proving the result. $\square$

Using the above lemma in (2.4.13), we obtain the error estimate

$$|e(x)| \leq \frac{1}{4(n+1)} \left| f^{(n+1)}(\xi) \right| h^{n+1} \leq \frac{M_{n+1}}{4(n+1)} h^{n+1}, \qquad (2.4.14)$$

where

$$M_{n+1} = \max_{\xi \in [a,b]} \left| f^{(n+1)}(\xi) \right|.$$

Example 2.5. Consider interpolating $f(x) = \sin(\pi x)$ with a polynomial, using uniformly distributed nodes on the interval $[-1, 1]$. Find an upper bound for error, and show how it is related with total number of nodes with some numerical simulations.

Answer. In this case we have $h = 2/n$, while

$$\left| f^{(n+1)}(\xi) \right| \leq \pi^{n+1} \qquad \text{for all } \xi.$$

Therefore, the upper bound for the error is

$$|e(x)| = |f(x) - P_n(x)| \leq \frac{\pi^{n+1}}{4(n+1)} \left(\frac{2}{n} \right)^{n+1}.$$

Below is a table of errors from simulations with various n.

n	error bound	measured error
4	4.8×10^{-1}	1.8×10^{-1}
8	3.2×10^{-3}	1.2×10^{-3}
16	1.8×10^{-9}	6.6×10^{-10}

We see that, although the error bounds are always bigger than the actual errors (they should be, if they are upper bounds), nevertheless, these bounds give the correct magnitude of the errors.

Matlab simulations. For the problem in Example 2.1, let us now plot the error $e(x) = f(x) - p_2(x)$ with (—) and upper error bound with (- - -), using the Matlab codes:

```
>> hold off
>> errorbound = abs(pi^3/48*t.*(t-1).*(t-2/3));
>> error = abs(cos(pi/2*t)-p2);
>> plot(t,error,t,errorbound,'--r')
```

The plot is shown in Figure 2.4. Notice how nicely the error bound dominates the actual error, with more or less the same magnitude.

If we increase the number of interpolation points by 1, choosing $n = 3$, we get the plots in Figure 2.5. We see that the error decreases as we increase n.

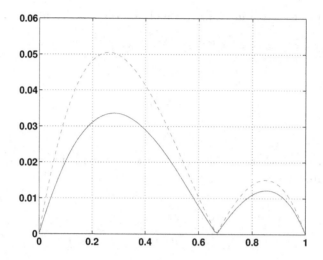

Fig. 2.4   Plot of the error and the error bound for Example 2.1.

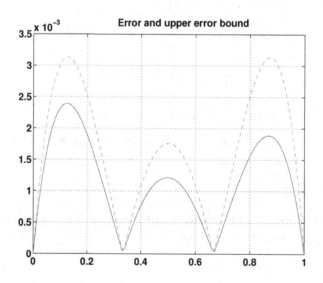

Fig. 2.5   Plots for Example 2.1 with $n = 3$.

As we further increase the interpolating points, we get the plots in Figure 2.6.

A problem encountered with uniformly spaced nodes is that the error can be large near the boundaries. This is revealed by a closer study of the Error Theorem 2.3, and is also shown by our numerical simulations. To avoid this problem, one could choose the interpolating points in a smarter way, i.e., not uniformly spaced

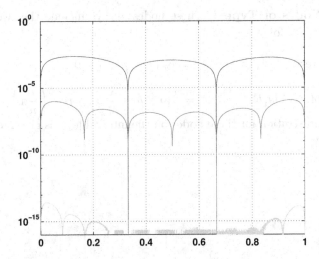

Fig. 2.6 Plot of the errors for Example 2.1 with $n = 3, 6, 12$.

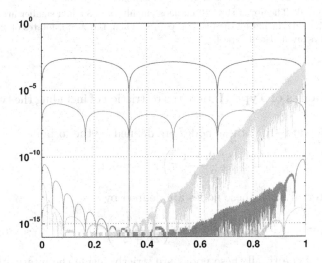

Fig. 2.7 Plot of the errors for Example 2.1 with $n = 3, 6, 12, 24, 48$.

in the interval $[a, b]$. This is the main idea in the construction of Chebyshev nodes.

Chebyshev nodes are a set of points on the interval $[a, b]$ such that the interpolating error has the same size on each sub-interval. To achieve this, one must choose a grid which is not uniform.

There are two types of Chebyshev nodes, one includes the endpoints a, b, the other does not. We shall look at both types.

Chebyshev Nodes of Type I: These nodes always include the two endpoints. They are defined as follows:

For the interval $[-1, 1]$: $\bar{x}_i = \cos\left(\dfrac{i}{n}\pi\right)$, $i = 0, 1, \cdots, n$,

For any interval $[a, b]$: $\bar{x}_i = \dfrac{1}{2}(a + b) + \dfrac{1}{2}(b - a)\cos\left(\dfrac{i}{n}\pi\right)$, $i = 0, 1, \cdots, n$.

A graphic illustration for these nodes on the interval $[a, b]$ is given in Figure 2.8.

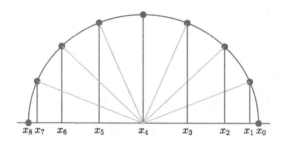

Fig. 2.8   Chebyshev nodes of type I, with $n = 8$. A half circle with radius $r = \frac{b-a}{2}$ is drawn, with the center at $(\frac{b+a}{2}, 0)$. The arc of the half circle is cut into $n = 8$ equal smaller arcs by $n + 1 = 9$ points on the half circle, including the two end points. Then, the x-coordinates of these points are exactly the Chebyshev nodes of type I.

Chebyshev Nodes of Type II: Here the restriction of including the two endpoints are removed.

For the interval $[-1, 1]$, these nodes are defined by the formula

$$\bar{x}_i = \cos\left(\frac{2i + 1}{2n + 2}\pi\right),   i = 0, 1, \cdots, n.$$

For any interval $[a, b]$, these nodes are defined by

$$\bar{x}_i = \frac{1}{2}(a + b) + \frac{1}{2}(b - a)\cos\left(\frac{2i + 1}{2n + 2}\pi\right),   i = 0, 1, \cdots, n.$$

Notice that for Type II, all these nodes are strictly inside the interval $[a, b]$. These nodes are roots of the Chebyshev polynomial of the first kind of degree n.

A graphic illustration for these nodes on the interval $[a, b]$ is given in Figure 2.9.

For Chebyshev nodes of type II on the interval $[-1, 1]$, one can show that

$$\max_{-1 \leq x \leq 1}\left\{\prod_{k=0}^{n}|x - \bar{x}_k|\right\} = 2^{-n} \leq \max_{-1 \leq x \leq 1}\left\{\prod_{k=0}^{n}|x - x_k|\right\}$$

where x_k is any other choice of nodes. This means, the term $\prod_{k=0}^{n}|x - \bar{x}_k|$, which appears in the Interpolation Error Theorem 2.3, is minimized by choosing the Chebyshev nodes as the interpolating points.

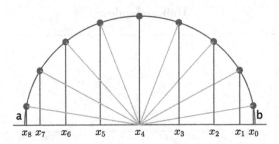

Fig. 2.9   Chebyshev nodes of type II, with $n = 8$. A half circle with radius $r = \frac{b-a}{2}$ is drawn, with the center at $(\frac{b+a}{2}, 0)$. The arc of half circle is cut into $n + 2 = 10$ small arcs by $n + 1 = 9$ points on the half circle. The middle $n = 8$ small arcs are equal to each other, while the left and right arcs are both half as long as the middle ones. Then, the x-coordinates of these points are exactly the Chebyshev nodes of type II.

This gives the following error bound:

$$|e(x)| \leq \frac{2^{-n}}{(n+1)!} \cdot \max_{\xi \in [-1,1]} \left| f^{(n+1)}(\xi) \right|.$$

Example 2.6. Consider the same example we used in uniform nodes example, i.e., Example 2.5, with $f(x) = \sin \pi x$. Now with Chebyshev nodes, we have

$$|e(x)| \leq \frac{2^{-n}}{(n+1)!} \pi^{n+1}.$$

Comparing these upper bounds with the actual errors we obtain the following table:

n	error bound	measured error
4	1.6×10^{-1}	1.15×10^{-1}
8	3.2×10^{-4}	2.6×10^{-4}
16	1.2×10^{-11}	1.1×10^{-11}

Notice that these errors are much smaller than those in Example 2.5 with a uniform grid!

Matlab Simulations. In Figure 2.10 we show plots of the interpolating polynomials and of the errors with uniform grid and with Chebyshev grid, taking $n = 4$. We show the same plots in Figure 2.10 with $n = 8$, and again the same plots in Figure 2.10 with $n = 16$. We see that the Chebyshev nodes are doing a much better job to decrease the error as the number of nodes doubles. Furthermore, the errors are evenly distributed throughout the interval, and no increase of the errors near the boundary points is observed.

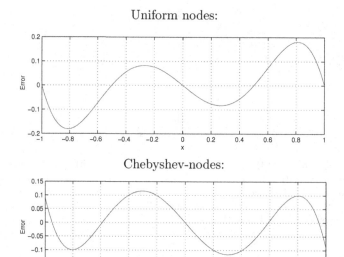

Fig. 2.10   Comparison of the interpolation errors with uniform and Chebyshev grid, for $n = 4$.

2.5   Properties of the Newton's Divided Differences (optional)

We now look at some interesting properties of Newton's divided differences.

Theorem 2.4. *If $P_n(x)$ interpolates $f(x)$ at $x_i \in [a, b]$, $i = 0, 1, \cdots, n$, then*

$$f(x) - P_n(x) = f[x_0, x_1, \cdots, x_n, x] \cdot \prod_{i=0}^{n}(x - x_i), \quad \text{for all } x \neq x_i.$$

Proof. Let $y \neq x_i$ for every $i = 0, \ldots, n$. Let $q(x)$ be the polynomial of degree $n + 1$ that interpolates $f(x)$ at $x_0, x_1, \cdots, x_n, y$. Newton's formula gives

$$q(x) = P_n(x) + f[x_0, x_1, \cdots, x_n, y] \prod_{i=0}^{n}(x - x_i).$$

Since $q(y) = f(y)$, we get

$$f(y) = q(y) = P_n(y) + f[x_0, x_1, \cdots, x_n, y] \prod_{i=0}^{n}(y - x_i).$$

Replacing y with x, we achieve the result. $\square$

As a consequence of the previous theorem we have:

Theorem 2.5. *For any set of distinct points $x_0, x_1, \ldots, x_n \in [a, b]$ there exists a value $\xi \in [a, b]$ such that*

$$f[x_0, x_1, \cdots, x_n] = \frac{1}{n!} f^{(n)}(\xi). \tag{2.5.15}$$

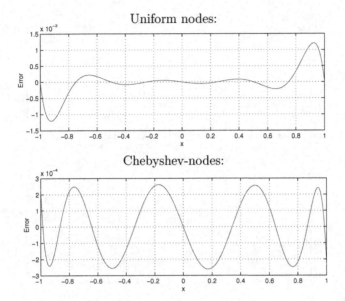

Fig. 2.11   Comparison of the interpolation errors with uniform and Chebyshev grid, for $n = 8$.

Proof. Let $P_{n-1}(x)$ interpolate $f(x)$ at $x_0, \cdots, x_{n-1}$. The error formula gives

$$f(x_n) - P_{n-1}(x_n) = \frac{1}{n!} f^{(n)}(\xi) \prod_{i=0}^{n} (x_n - x_i), \qquad \xi \in (a, b).$$

From the above we know

$$f(x_n) - P_{n-1}(x_n) = f[x_0, \cdots, x_n] \prod_{i=0}^{n} (x_n - x_i).$$

Comparing the right-hand sides of these two equations, we get the result. $\square$

Observation: Newton's divided differences are related to derivatives.

When $n = 1$, choosing $x_0 = x$ and $x_1 = x + h$, we obtain

$$f[x_0, x_1] = \frac{f(x + h) - f(x)}{h} = f'(\xi) \qquad \text{for some } \xi \in [x_0, x_1].$$

When $n = 2$, choosing $x_0 = x - h$, $x_1 = x$, and $x_2 = x + h$, we obtain

$$f[x_0, x_1, x_2] = \frac{1}{2h^2}[f(x+h) - 2f(x) + f(x+h)] = \frac{1}{2}f''(\xi) \qquad \text{for some } \xi \in [x_0, x_1].$$

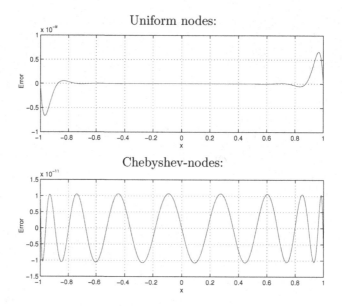

Fig. 2.12   Comparison of the interpolation errors with uniform and Chebyshev grid, for $n = 16$.

2.6   Convergence of Polynomial Interpolation

Here we briefly discuss the convergence issue. The main question is: As $n \to +\infty$, does the polynomial $P_n(x)$ converge to the function $f(x)$?

Convergence may be understood in different ways. Let $e_n(x) = f(x) - P_n(x)$ be the error function. We list the main types of convergence:

$$\text{uniform convergence}: \quad \lim_{n \to +\infty} \max_{a \le x \le b} |e_n(x)| = 0.$$

$$\mathbf{L}^1 \text{ convergence}: \quad \lim_{n \to +\infty} \int_a^b |e_n(x)| \, dx = 0.$$

$$\mathbf{L}^2 \text{ convergence}: \quad \lim_{n \to +\infty} \int_a^b |e_n(x)|^2 \, dx = 0.$$

For uniformly distributed nodes, it is known that for some functions $f(x)$, the error can become arbitrarily large as $n \to +\infty$. We have observed this in our simulations, and it is very bad news!

It is possible to achieve convergence, however. For each function $f(x)$, there is a way of choosing a sequence of nodes $\{x_i\}_{i=0}^n$, such that $|e_n(x)| \to 0$ as $n \to +\infty$. However, the sequence of points x_i must be carefully chosen, and can depend on the function f. With the wrong sequence of interpolating points, there will be no convergence. This is not practical!

Conclusion. One may reach the conclusion that, if one wishes to represent a function or a data set with polynomials, then increasing the order of the polynomial might not be the best thing to do. It is better to subdivide the interval $[a, b]$ into many small sub-intervals, and approximate the function with a polynomial of low order, on each subinterval. This is precisely the idea that will be developed in the next chapter.

2.7   Homework Problems for Chapter 2

NB!   If a problem does not ask you to use Matlab, then you should not use Matlab. You may use Matlab as a fancy calculator.

Problem 1

(a).   Use the Lagrange interpolation process to obtain a polynomial of least degree that assumes these values:

x	0	2	3	4
y	7	11	28	63

(b).   For the points in the Table of (a), find the Newton's form of the interpolating polynomial. Show that the two polynomials obtained are identical, although their forms may differ.

(c).   The polynomial $p(x) = x^4 - x^3 + x^2 - x + 1$ has the values shown.

x	-2	-1	0	1	2	3
$p(x)$	31	5	1	1	11	61

Find a polynomial $q(x)$ that takes these values (you don't need expand it):

x	-2	-1	0	1	2	3
$q(x)$	31	5	1	1	11	30

(Hint: This can be done with little work. Try the Lagrange form.)

Problem 2

(a).   Show directly that the maximum error associated with linear interpolation of a function $f(x)$ at the points x_0, x_1, is bounded by $\frac{1}{8}(x_1 - x_0)^2 M$, where $M = \max_{x_0 \le x \le x_1} |f''(x)|$.

(b).   An interpolating polynomial of degree 20 is to be used to approximate e^{-x} on the interval $[0, 2]$. Give a rough upper bound for the error.

Problem 3: A Study on Polynomial Interpolation in Matlab

The goal of this exercise is to get more familiar with Matlab.

The target problem: Find the interpolating polynomial
$$p_3(x) = a_3 x^3 + a_2 x^2 + a_1 x + a_0 \tag{2.7.16}$$
that interpolates the following data points:

x_i	0	5	10	15
y_i	3	8	-2	9

$$\tag{2.7.17}$$

Preparations:

a) Show that the coefficients to the interpolating polynomial (2.7.16) can be found by solving

$$\begin{pmatrix} 1 & 0 & 0 & 0 \\ 1 & 5 & 25 & 125 \\ 1 & 10 & 100 & 1000 \\ 1 & 15 & 225 & 3375 \end{pmatrix} \begin{pmatrix} a_0 \\ a_1 \\ a_2 \\ a_3 \end{pmatrix} = \begin{pmatrix} 3 \\ 8 \\ -2 \\ 9 \end{pmatrix} \qquad (2.7.18)$$

b) Read through Chapter 3 "Programming in Matlab" in "*A Practical Introduction to Matlab*" by Gockenbach, at the web-page
   http://www.math.mtu.edu/~msgocken/intro/intro.pdf

Linear equations
Solve the system of linear equations (2.7.18) using Matlab. The interpolating polynomial should be

$$p_3(x) = 0.048x^3 - 1.02x^2 + 4.9x + 3.$$

Simple plots
Plot the interpolating points and the interpolating polynomial. In addition to the function `plot` try also:

   `grid on/off`: add or remove a grid on the plot.
   `xlabel('text')`: put text under x-axis.
   `ylabel('text')`: put text next to y-axis.
   `title('text')`: put text above the plot.

Helpdesk. If you want to find out what Matlab has to offer in connection with polynomials, you may type in the command:

   `lookfor polynomial`

and see what pops up. Check out how to use the functions `polyfit` and `polyval`. For example, `help polyfit`, would give you the following information:

```
POLYFIT Fit polynomial to data.
    POLYFIT(X,Y,N) finds the coefficients of a polynomial P(X) of
    degree N that fits the data, P(X(I))~=Y(I), in a least-squares
    sense.  ...
```

Here X is a vector with the x-values, and Y is a vector with y-values from the table in (2.7.17) in our example. N is the order of the polynomial, and in our case we use $N = 3$. The function returns a vector with the coefficients of the polynomial.

   Use these functions `polyfit` and `polyval` to compute and plot the interpolating polynomial for (2.7.17).

Problem 4

Let the function $f(x)$ be three times differentiable. Consider the finite difference approximation formula

$$f'(x) \approx D_h(x) = \frac{1}{2h}[-3f(x) + 4f(x+h) - f(x+2h)]. \qquad (2.7.19)$$

Note that this scheme uses values of f at the three points $x, x+h, x+2h$. This is a one-sided finite difference.

(a). Using Taylor series, show that the local truncation error is bounded by Ch^2 for some constant C, i.e.

$$|f'(x) - D_h(x)| \leq Ch^2.$$

(b). We now sharpen the error estimate in (a). Show that the local truncation error is

$$|f'(x) - D_h(x)| = \frac{1}{3}h^2 f'''(\xi), \qquad \text{for some} \quad \xi \in [x, x+2h].$$

(Hints: Consider the polynomial that interpolates the function f at the following 3 points: $x, x+h, x+2h$.)

(c). Now set $f(x) = \tan x$. Use the formula in (2.7.19) to approximate the derivative $f'(x)$ at $x = 1.0$, choosing $h = 0.1$, then $h = 0.01$ and $h = 0.001$. Compare with the exact value to get the errors. Comment on your results. Are these errors consistent with the error estimate in (a)?

(d). Use Matlab to compute the approximation of the derivative of $\tan x$ at $x = 1.0$, with the formula in (2.7.19), with smaller and smaller values of h. Plot the error e as a function of h, use log scale (see Matlab command `loglog`). How did you expect the error to behave? How does the actual error behave? For what h value do you get the best result? What do you think is the cause of this behavior?

   What to hand in for part (d): Hand in the Matlab script file, the relevant output of your code, the plots of errors, and your comments.

Problem 5: Newton's Divided Difference in Matlab

(a). Write two functions in Matlab. The first function, called `divdiff`, should read in two vectors x and y, and return a table (a matrix) of the divided difference values. This means, the first few lines of your `divdiff.m` file should be:

```
function a=divdiff(x,y)
% input:  x,y: the data set to be interpolated
% output: a: table for Newton's divided differences.
```

   The second function, called `polyvalue`, should read in the table of divided difference values generated by the function `divdiff`, the x-vector, and a vector t.

The output of the function are the values of Newton's polynomial computed at points given in the vector t. This means, your file `polyvalue.m` should start with the following few lines:

```
function v=polyvalue(a,x,t)
% input: a= Newton's divided differences
%        x= the points for the data set to interpolate,
%                same as in divdiff.
%        t= the points where the polynomial should be evaluated
% output: v= value of polynomial at the points in t
```

What to hand in: Hand in the files `divdiff.m` and `polyvalue.m`.

(b). Here you can test your Matlab functions in (a). Using 21 equally spaced nodes on the interval $[-5, 5]$, find the interpolating polynomial p of degree 20 for the function $f(x) = (x^2 + 1)^{-1}$. Plot the functions $f(x)$ and $p(x)$ together at 41 equally spaced points, including the nodes. Plot also the error $e(x) = |f(x) - p(x)|$. Why does this work so poorly? Write your comments.

What to hand in: Hand in your Matlab script file, and the plots of $f(x)$ with $p(x)$, and the error, and your comments.

(c). With the same problems in (b), now we use the Chebyshev nodes of Type I and Type II:

$$\text{Type I:} \qquad x_i = 5\cos(i\pi/20), \qquad 0 \le i \le 20,$$

$$\text{Type II:} \qquad x_i = 5\cos[(2i + 1)\pi/42], \qquad 0 \le i \le 20.$$

Find the interpolating polynomial $p(x)$ for each Type, and plot $f(x)$ with $p(x)$ for both Types, and the errors $e(x) = |f(x) - p(x)|$ as well. Do these nodes work better than uniform nodes? Give your comments.

What to hand in: Hand in your Matlab script file, and the plots of $f(x)$ with $p(x)$, and the error, and your comments.

Chapter 3

Piecewise Polynomial Interpolation. Splines

3.1 Introduction

As we mentioned at the end of last chapter, because of the problems encountered with polynomial interpolation, it is better to use piecewise polynomial interpolation. In other words, instead of approximating a function f with a single polynomial of high degree over the entire interval $[a, b]$, it is better to use different polynomials of lower degree, separately on different subintervals. Such functions are called *spline functions*.

Spline interpolations are similar to polynomial interpolation, possibly with lower smoothness (i.e., differentiability) requirements. For example:

- Visualization of discrete data: the plot needs only to "look smooth" to human eyes.
- Graphic design: the curve or surface needs to go through certain control points, and look smooth, with less curvature.

Some basic requirements of splines interpolations are:

- Correct interpolation of data points.
- Some degree of smoothness, usually not very high.
- Convergence to the original function f, as the number of data points increases.

We recall again some disadvantages of polynomial interpolation discussed in the previous chapter:

- The interpolating polynomials are n-times differentiable. We do not need such high smoothness.
- The error is big in certain regions, especially near the endpoints.
- As $n \to \infty$, there is no guarantee of convergence.
- Such polynomials are expensive to compute, for large n.

In order to achieve the above requirements and avoid the difficulties encountered with polynomial interpolation, we shall subdivide the interval $[a, b]$ into smaller sub-

47

intervals, and use piecewise polynomial interpolation separately on each subinterval, where each polynomial is of low degree. Normally the degree of these polynomials will be ≤ 3.

Problem setting: Given a set of data

$$
\begin{array}{c|cccc}
x & t_0 & t_1 & \cdots & t_n \\
\hline
y & y_0 & y_1 & \cdots & y_n
\end{array},
$$

we want to find a piecewise polynomial function $S(x)$ which interpolates the points $(t_i, y_i)_{i=0}^n$.

The points $t_0 < t_1 < \cdots < t_n$ are called *knots*. Note that they need to be given in *increasing order*.

The interpolating spline function $S(x)$ consists of piecewise polynomials in the following way

$$
S(x) \doteq \begin{cases}
S_0(x), & t_0 \leq x \leq t_1, \\
S_1(x), & t_1 \leq x \leq t_2, \\
\vdots & \\
S_{n-1}(x), & t_{n-1} \leq x \leq t_n.
\end{cases} \tag{3.1.1}
$$

Below is the definition of a spline of degree k.

Definition 3.1. The piecewise polynomial function $S(x)$ in (3.1.1) is called a *spline of degree k* if the following holds.

- Each $S_i(x)$ is a polynomial of degree $\leq k$;
- $S(x)$ is $(k-1)$ times continuously differentiable. Namely, at every intermediate knot t_i, with $i = 1, 2, \cdots, n-1$, we have

$$
\begin{aligned}
S_{i-1}(t_i) &= S_i(t_i), \\
S'_{i-1}(t_i) &= S'_i(t_i),
\end{aligned}
$$

$$
\vdots
$$

$$
S_{i-1}^{(k-1)}(t_i) = S_i^{(k-1)}(t_i).
$$

Commonly used splines include the lowest degree ones, i.e.,

- $k = 1$: linear spline (simplest)
- $k = 2$: quadratic spline (less popular)
- $k = 3$: cubic spline (most popular, for good reasons)

Splines with degree larger than 3 are rarely used.

We now consider several examples, checking if a given piecewise polynomial function is a spline function, and determining its degree. In each case, we check whether all the conditions in Definition 3.1 are satisfied.

Example 3.1. Determine whether this function is a linear spline (i.e., a spline function of degree 1):

$$S(x) = \begin{cases} x, & x \in [-1,0], \\ 1-x, & x \in (0,1), \\ 2x-2, & x \in [1,2]. \end{cases}$$

Answer. We now check all the properties of a linear spline.

- $S(x)$ is a linear polynomial, on each subinterval. TRUE
- $S(x)$ is continuous at inner knots. FALSE: At $x = 0$, $S(x)$ is discontinuous, because from the left we get the limit 0 and from the right we get the limit 1.

Therefore this is NOT a linear spline.

Example 3.2. Determine whether the following function is a quadratic spline (i.e., a spline function of degree 2):

$$S(x) = \begin{cases} x^2, & x \in [-10,0], \\ -x^2, & x \in (0,1), \\ 1-2x, & x \geq 1. \end{cases}$$

Answer. Let us label each polynomial as

$$Q_0(x) = x^2, \quad Q_1(x) = -x^2, \quad Q_2(x) = 1 - 2x.$$

Clearly, these are all polynomials of degree 2. We now check all the remaining conditions, i.e., the continuity of Q and Q' at the inner knots $0, 1$:

$$\begin{aligned} Q_0(0) &= 0 = Q_1(0), \\ Q_1(1) &= -1 = Q_2(1), \\ Q_0'(0) &= 0 = Q_1'(0), \\ Q_1'(1) &= -2 = Q_2'(1). \end{aligned}$$

It passes all the tests. Therefore, $S(x)$ is indeed a quadratic spline.

3.2 Linear Splines

We consider here the case where $k = 1$. This amounts to a piecewise linear interpolation, where we simply connect 2 neighboring points with the straight line between them. This is called a *linear spline function*. See Figure 3.1 for an illustration.

On each subinterval we thus have a linear polynomial, which can be written as

$$S_i(x) = a_i + b_i x, \quad i = 0, 1, \cdots, n-1.$$

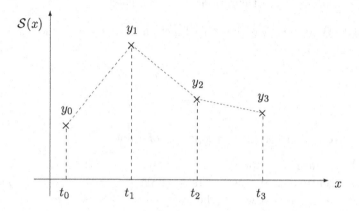

Fig. 3.1   Linear splines

According to the definition of a spline with degree $k = 1$, we have the following requirements:

$$\mathcal{S}_0(t_0) = y_0,$$
$$\mathcal{S}_{i-1}(t_i) = \mathcal{S}_i(t_i) = y_i, \qquad i = 1, 2, \cdots, n-1,$$
$$\mathcal{S}_{n-1}(t_n) = y_n.$$

These linear polynomials are easy to compute. Indeed, for every $i = 0, \ldots, n-1$, the equation of the line passing through the two points (t_i, y_i) and (t_{i+1}, y_{i+1}) is given by

$$\mathcal{S}_i(x) = y_i + \frac{y_{i+1} - y_i}{t_{i+1} - t_i}(x - t_i), \qquad i = 0, 1, \cdots, n-1.$$

We state an Error Theorem for linear spline interpolation, without proof.

Theorem 3.1. *(Accuracy of a linear spline)* *Consider the points*

$$t_0 < t_1 < t_2 < \cdots < t_n,$$

and let

$$h_i = t_{i+1} - t_i, \qquad h = \max_{0 \le i \le n-1} h_i.$$

Let $f(x)$ be a given function and let $\mathcal{S}(x)$ be a linear spline that interpolates $f(x)$ at the knots t_i, so that

$$\mathcal{S}(t_i) = f(t_i), \qquad \text{for all } i = 0, 1, \cdots, n.$$

Then, for every $x \in [t_0, t_n]$, one has the following error estimates.

(i) If f' is continuous, then

$$|f(x) - \mathcal{S}(x)| \le \max_i \left\{ \frac{1}{2} h_i \cdot \max_{t_i \le \xi \le t_{i+1}} |f'(\xi)| \right\} \le \frac{h}{2} \max_{t_0 \le \xi \le t_n} |f'(\xi)|.$$

(ii) If f'' is continuous, then

$$|f(x) - S(x)| \leq \max_i \left\{ \frac{1}{8} h_i^2 \cdot \max_{t_i \leq \xi \leq t_{i+1}} |f''(\xi)| \right\} \leq \frac{h^2}{8} \max_{t_0 \leq \xi \leq t_n} |f''(\xi)|.$$

To minimize error, it is clear that one should add more knots where the function has large first or second derivatives, i.e., where the function $f(x)$ changes quickly.

3.3 Quadratic Splines (optional)

This type of splines is not much used. As you will see, the computation of such a spline function involves quite some work. In return, the performance of quadratic splines is poorer than those of cubic splines. In practice, cubic splines are usually favored for their minimum curvature property (this will be discussed in detail in the next section).

Given a set of knots $t_0, t_1, \cdots, t_n$, and data values $y_0, y_1, \cdots, y_n$, we seek a piecewise polynomial representation

$$Q(x) = \begin{cases} Q_0(x), & t_0 \leq x \leq t_1, \\ Q_2(x), & t_1 \leq x \leq t_2, \\ \vdots \\ Q_{n-1}(x), & t_{n-1} \leq x \leq t_n, \end{cases}$$

where each $Q_i(x)$ is a quadratic polynomial. In general, we can write

$$Q_i(x) = a_i x^2 + b_i x + c_i.$$

We need to find the coefficients a_i, b_i, c_i for $i = 0, 1, \cdots, n - 1$. The total number of unknowns is $3n$.

On each Q_i we impose the following conditions:

$$Q_i(t_i) = y_i, \qquad \text{for } i = 0, 1, \cdots, n - 1: \qquad n \text{ conditions,}$$

$$Q_i(t_{i+1}) = y_{i+1}, \qquad \text{for } i = 0, 1, \cdots, n - 1: \qquad n \text{ conditions,}$$

$$Q_i'(t_i) = Q_{i+1}'(t_i), \qquad \text{for } i = 1, 2, \cdots, n - 1: \qquad n - 1 \text{ conditions.}$$

Total number of conditions we have so far is $2n + (n - 1) = 3n - 1$.

Since we have $3n$ unknowns, to achieve a unique solution we must impose one more condition. This additional condition is our free choice. For example, we may require $Q_0'(t_0) = 0$, or $Q_0''(t_0) = 0$, depending on the specific problem.

Construction of $Q_i(t)$: Since $Q(x)$ is a quadratic spline, then Q' is a linear spline, and it is continuous. We denote

$$z_i = Q'(t_i).$$

At this stage, these values z_i are still unknown, and will be computed later.

We observe that, as soon as the z_i are found, one can easily construct the quadratic splines Q_i. Indeed, each Q_i must satisfy the conditions:

$$Q_i(t_i) = y_i, \tag{3.3.2}$$

$$Q_i(t_{i+1}) = y_{i+1}, \tag{3.3.3}$$

$$Q_i'(t_i) = z_i, \tag{3.3.4}$$

$$Q_i'(t_{i+1}) = z_{i+1}. \tag{3.3.5}$$

For every i, using the three conditions (3.3.2), (3.3.4) and (3.3.5), we obtain the polynomial

$$Q_i(x) = \frac{z_{i+1} - z_i}{2(t_{i+1} - t_i)}(x - t_i)^2 + z_i(x - t_i) + y_i, \qquad 0 \le i \le n - 1. \tag{3.3.6}$$

It is easy to verify this polynomial satisfies the above three conditions (but not necessarily the remaining condition (3.3.3) at the point t_{i+1}).

To find the values for z_i, we now use the identity (3.3.3). This yields

$$z_{i+1} = -z_i + 2\left(\frac{y_{i+1} - y_i}{t_{i+1} - t_i}\right), \qquad 0 \le i \le n - 1. \tag{3.3.7}$$

Given z_0, we can compute z_1 using (3.3.7) with $i = 0$. Then we use z_1 to compute z_2, etc. By induction, all the values z_i can be computed.

Once the z_i's are known, we obtain the polynomials Q_i using the formulas (3.3.6).

The algorithm is summarized below.

Quadratic Spline Algorithm:

- Given z_0, compute z_i using (3.3.7).
- Compute Q_i by using (3.3.6).

3.4   Natural Cubic Splines

Given $t_0 < t_1 < \cdots < t_n$, we define the *cubic spline*

$$S(x) = S_i(x), \qquad \text{for} \quad t_i \le x \le t_{i+1}, \quad 0 \le i \le n - 1.$$

By definition of a cubic spline, we require that S, S', S'' are all continuous.

Moreover, each S_i is a cubic polynomial, so that

$$S_i(x) = a_i x^3 + b_i x^2 + c_i x + d_i, \qquad i = 0, 1, \cdots, n - 1.$$

Here the total number of unknowns is $4 \cdot n$.

We now collect all the requirements in the definition of a cubic spline function:

Equations   #eqns

(1) $S_i(t_i) = y_i,$	$i = 0, 1, \cdots, n-1$	(n)
(2) $S_i(t_{i+1}) = y_{i+1},$	$i = 0, 1, \cdots, n-1$	(n)
(3) $S_i'(t_{i+1}) = S_{i+1}'(t_{i+1}),$	$i = 0, 1, \cdots, n-2$	$(n-1)$
(4) $S_i''(t_{i+1}) = S_{i+1}''(t_{i+1}),$	$i = 0, 1, \cdots, n-2$	$(n-1)$
(5) $S_0''(t_0) = 0,$		1
(6) $S_{n-1}''(t_n) = 0,$		1

Note that the total number of equations without (5)-(6) equals to $4n - 2$, which is smaller than the number of unknowns by 2. Therefore the above equations (5) and (6) above are additional requirements that we impose, so that the number of equations is equal to the number of unknowns. When the extra conditions $S_0''(t_0) = S_{n-1}''(t_n) = 0$, are imposed, the cubic spline is called a *natural cubic spline*.

Of course there are other extra conditions one can use, more suitable for specific problems. These would lead to different types of cubic splines.

Let us see how to compute $S_i(x)$. By assumption, we know that

S_i is a polynomial of degree 3,
S_i' is a polynomial of degree 2,
S_i'' is a polynomial of degree 1.

Here is the overview of the solution procedure:

- We start with $S_i''(x)$. They are all linear polynomials, and one can write them out in Lagrange form.
- We then integrate $S_i''(x)$ twice to get an expression for $S_i(x)$. For this purpose, two integration constants must be found.
- These integration constants can be determined using the equations (2) and (1) in the previous formula. There are various tricks to compute these constants in an efficient way.

We now describe the details of this construction. We start with S'' and define z_i as

$$z_i = S''(t_i), \qquad i = 1, 2, \cdots, n-1, \qquad z_0 = z_n = 0.$$

Note that these z_i's are our main unknowns!

We introduce the notation

$$h_i = t_{i+1} - t_i.$$

The linear polynomials S_i'' can be written in Lagrange form as

$$S_i''(x) = \frac{z_{i+1}}{h_i}(x - t_i) - \frac{z_i}{h_i}(x - t_{i+1}). \tag{3.4.8}$$

We integrate twice, and write the anti-derivatives (note that it is very useful to write them in these particular forms!):

$$S_i'(x) = \frac{z_{i+1}}{2h_i}(x - t_i)^2 - \frac{z_i}{2h_i}(x - t_{i+1})^2 + C_i - D_i,$$

$$S_i(x) = \frac{z_{i+1}}{6h_i}(x - t_i)^3 - \frac{z_i}{6h_i}(x - t_{i+1})^3 + C_i(x - t_i) - D_i(x - t_{i+1}).$$

By a straightforward differentiation, one can check that these S_i, S_i' are the correct anti-derivatives for S_i'' in (3.4.8), for any choice of the constants C_i, D_i. We skip the details here and ask the students to fill in the missing computations.

We now impose the interpolation properties.

- The condition $S_i(t_i) = y_i$ yields

$$y_i = -\frac{z_i}{6h_i}(-h_i)^3 - D_i(-h_i) = \frac{1}{6}z_i h_i^2 + D_i h_i.$$

This gives us a formula to compute D_i, namely

$$D_i = \frac{y_i}{h_i} - \frac{h_i}{6}z_i.$$

- The condition $S_i(t_{i+1}) = y_{i+1}$ yields

$$y_{i+1} = \frac{z_{i+1}}{6h_i}h_i^3 + C_i h_i.$$

This gives us a formula to compute C_i, namely

$$C_i = \frac{y_{i+1}}{h_i} - \frac{h_i}{6}z_{i+1}.$$

We see that, if the z_i are known, then the values C_i, D_i are known, and hence we can determine S_i, S_i'. Indeed, we have

$$S_i(x) = \frac{z_{i+1}}{6h_i}(x - t_i)^3 - \frac{z_i}{6h_i}(x - t_{i+1})^3 + \left(\frac{y_{i+1}}{h_i} - \frac{h_i}{6}z_{i+1}\right)(x - t_i)$$

$$- \left(\frac{y_i}{h_i} - \frac{h_i}{6}z_i\right)(x - t_{i+1}). \tag{3.4.9}$$

$$S_i'(x) = \frac{z_{i+1}}{2h_i}(x - t_i)^2 - \frac{z_i}{2h_i}(x - t_{i+1})^2 + \frac{y_{i+1} - y_i}{h_i} - \frac{z_{i+1} - z_i}{6}h_i. \tag{3.4.10}$$

Finally, we need to compute the z_i's. The remaining condition, which has not yet been used, is the continuity of $S'(x)$, i.e.,

$$S_{i-1}'(t_i) = S_i'(t_i), \qquad i = 1, 2, \cdots, n - 1.$$

These conditions yield

$$S_i'(t_i) = -\frac{z_i}{2h_i}(-h_i)^2 + \underbrace{\frac{y_{i+1} - y_i}{h_i} - \frac{z_{i+1} - z_i}{6}h_i}_{b_i}$$

$$= -\frac{1}{6}h_i z_{i+1} - \frac{1}{3}h_i z_i + b_i,$$

$$S_{i-1}'(t_i) = \cdots = \frac{1}{6}z_{i-1}h_{i-1} + \frac{1}{3}z_i h_{i-1} + b_{i-1}.$$

Setting $S_{i-1}'(t_i) = S_i'(t_i)$ and multiplying the resulting equation by 6, we obtain

$$\begin{cases} h_{i-1}z_{i-1} + 2(h_{i-1} + h_i)z_i + h_i z_{i+1} = 6(b_i - b_{i-1}), & i = 1, 2, \cdots, n-1, \\ z_0 = 0, \quad z_n = 0. \end{cases}$$

$$(3.4.11)$$

Note that the two boundary values $z_0 = z_n = 0$ come from the additional conditions (5) and (6) for the *natural* cubic spline.

We can write the system of $n-1$ equations (3.4.11) in the following matrix-vector form:

$$\mathbf{H} \cdot \vec{z} = \vec{v} \qquad (3.4.12)$$

where

$$\mathbf{H} = \begin{pmatrix} 2(h_0 + h_1) & h_1 & & & & \\ h_1 & 2(h_1 + h_2) & h_2 & & & \\ & h_2 & 2(h_2 + h_3) & h_3 & & \\ & & \ddots & \ddots & \ddots & \\ & & & h_{n-3} & 2(h_{n-3} + h_{n-2}) & h_{n-2} \\ & & & & h_{n-2} & 2(h_{n-2} + h_{n-1}) \end{pmatrix}$$

and

$$\vec{z} = \begin{pmatrix} z_1 \\ z_2 \\ z_3 \\ \vdots \\ z_{n-2} \\ z_{n-1} \end{pmatrix}, \qquad \vec{v} = \begin{pmatrix} 6(b_1 - b_0) \\ 6(b_2 - b_1) \\ 6(b_3 - b_2) \\ \vdots \\ 6(b_{n-2} - b_{n-3}) \\ 6(b_{n-1} - b_{n-2}) \end{pmatrix}.$$

We observe that $\mathbf{H}$ is a tridiagonal symmetric matrix. Moreover, it is diagonally dominant because

$$2|h_{i-1} + h_i| > |h_i| + |h_{i-1}|.$$

Therefore, the vector $\vec{z}$ can be determined as the unique solution of the system (3.4.12).

We can now summarize the algorithm.

Natural Cubic Spline Algorithm:

- Find the coefficients z_i, solving the linear system of equations (3.4.12);
- Compute $S_i(x)$ using (3.4.9).

Matlab Simulations. The following Matlab function will set up the tridiagonal system (3.4.12), using 2 vectors to store the information for the matrix **H**. Among these 2 vectors, one of them represents the diagonal elements, and the other represents both the upper and lower diagonal elements since the coefficient matrix is symmetric. Then a simple Gaussian Elimination procedure, especially designed for this tridiagonal system, is applied to solve the system (3.4.12).

```
function z = cspline(t,y)
n = length(t);
z = zeros(n,1);
h = zeros(n-1,1);
b = zeros(n-1,1);
u = zeros(n,1);
v = zeros(n,1);

h = t(2:n)-t(1:n-1);
b = (y(2:n)-y(1:n-1))./h;

u(2) = 2*(h(1)+h(2));   % diagonal of H matrix
v(2) = 6*(b(2)-b(1));   % the right-hand side vector

% Gaussian Elimination.  Forward Elimination
for i=3:n-1
  u(i) = 2*(h(i)+h(i-1))-h(i-1)^2/u(i-1);
   v(i) = 6*(b(i)-b(i-1))-h(i-1)*v(i-1)/u(i-1);
end

% Backward substitution
for i=n-1:-1:2
   z(i) = (v(i)-h(i)*z(i+1))/u(i);
end
```

For any given value x, we can compute the value $S(x)$ using the following code:

```
function S = cspline_eval(t,y,z,x)
m = length(x);
```

```
n = length(t);
for i=n-1:-1:1
  if (x-t(i)) >= 0
    break
  end
end
h = t(i+1)-t(i);
S = z(i+1)/(6*h)*(x-t(i))^3 ...
   -z(i)/(6*h)*(x-t(i+1))^3 ...
   +(y(i+1)/h-z(i+1)*h/6)*(x-t(i)) ...
   -(y(i)/h-z(i)*h/6)*(x-t(i+1));
```

These Matlab functions can be used in the following way, in the command window of Matlab:

```
>> t = [0.9,1.3,1.9,2.1]
t =
     0.9000    1.3000    1.9000    2.1000
>> y = [1.3,1.5,1.85,2.1]
y =
     1.3000    1.5000    1.8500    2.1000
>> z = cspline(t,y)
z =
          0
    -0.5634
     2.7113
          0
>> cspline_eval(t,y,z,1.5)
ans =
     1.5810
```

If one wants to compute the values of $S(x)$ where x is a vector of many x-values, then the function cspline_eval shall be slightly modified.

```
function S = cspline_eval(t,y,z,x_vec)
% function S = cspline_eval(t,y,z,x_vec)
% compute the value of the natural cubic spline
% at the points x_vec when t,y,z are given
%
% Example:   t = [0.9,1.3,1.9,2.1];
%            y = [1.3,1.5,1.85,2.1]
%            z = cspline(t,y)
```

```
%                x = [0.9:0.1:2.1]
%                v = cspline_eval(t,y,z,x)

m = length(x_vec);
S = zeros(size(x_vec));
n = length(t);
for j=1:m
  x = x_vec(j);
  for i=n-1:-1:1
    if (x-t(i)) >= 0
      break
    end
  end
  h = t(i+1)-t(i);
  S(j) = z(i+1)/(6*h)*(x-t(i))^3-z(i)/(6*h)*(x-t(i+1))^3 ...
         +(y(i+1)/h-z(i+1)*h/6)*(x-t(i)) ...
         - (y(i)/h-z(i)*h/6)*(x-t(i+1));
end
```

Comparison between polynomial interpolation and natural cubic spline interpolation. We now perform a comparison between polynomial interpolation and the natural cubic spline, for the same given set of data. The result is in Figure 3.2, where

- The markers 'o': interpolating points or knots;
- The left plot: polynomial interpolation;
- The right plot: Natural cubic spline interpolation.

We observe that the polynomial interpolation has an oscillatory behavior near the two end points. It is clear that the natural cubic spline gives a smoother profile. In a way, the natural cubic spline provides the "best possible" interpolation, as stated in the following famous Theorem.

Theorem 3.2. *(Smoothness of Natural Cubic Splines) Let S be the natural cubic spline function that interpolates a twice-continuously differentiable function f at knots*

$$a = t_0 < t_1 < \cdots < t_n = b.$$

Then we have

$$\int_a^b [S''(x)]^2 \, dx \le \int_a^b [f''(x)]^2 \, dx.$$

Note that $\int (f''(x))^2 \, dx$ is related to the curvature of the graph of the function f. Intuitively, this theorem states that, given a fixed set of knots, the natural cubic

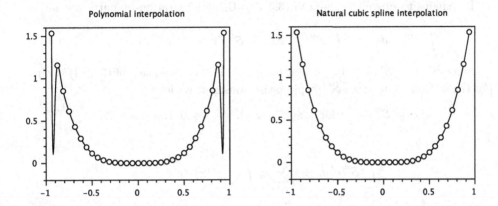

Fig. 3.2 Comparison between polynomial interpolation and the natural cubic spline.

spline gives the least curvature among all functions that interpolates the same data. Therefore, the natural cubic spline is your best choice, if you want a smooth profile.

Proof. Consider the error function

$$g(x) = f(x) - \mathcal{S}(x).$$

Then, we have

$$g(t_i) = 0, \qquad i = 0, 1, \cdots, n.$$

Differentiating twice, we have

$$f'' = \mathcal{S}'' + g'',$$

hence

$$(f'')^2 = (\mathcal{S}'')^2 + (g'')^2 + 2\mathcal{S}''g''.$$

Integrating both sides, we get

$$\int_a^b (f'')^2 \, dx = \int_a^b (\mathcal{S}'')^2 \, dx + \int_a^b (g'')^2 \, dx + \int_a^b 2\mathcal{S}''g'' \, dx.$$

We now **claim** that

$$\int_a^b \mathcal{S}''g'' \, dx = 0, \tag{3.4.13}$$

then this would imply

$$\int_a^b (f'')^2 \, dx \geq \int_a^b (\mathcal{S}'')^2 \, dx,$$

and we are done.

It remains to prove the claim in (3.4.13). Using integration-by-parts, we get

$$\int_a^b \mathcal{S}'' g'' \, dx \;=\; \int_a^b \mathcal{S}'' (g')' \, dx \;=\; \mathcal{S}'' g' \Big|_a^b - \int_a^b \mathcal{S}''' g' \, dx. \qquad (3.4.14)$$

Since $\mathcal{S}''(a) = \mathcal{S}''(b) = 0$, the first term on the right-hand side of (3.4.14) is zero. For the second term, since $\mathcal{S}'''$ is piecewise constant, we let

$$c_i = \mathcal{S}'''(x), \quad \text{for} \quad x \in [t_i, t_{i+1}], \qquad i = 0, 1, \cdots, n-1.$$

Then

$$\int_a^b \mathcal{S}''' g' \, dx = \sum_{i=0}^{n-1} c_i \int_{t_i}^{t_{i+1}} g'(x) \, dx$$

$$= \sum_{i=0}^{n-1} c_i \left[g(t_{i+1}) - g(t_i) \right] = 0,$$

because $g(t_i) = 0$ for every i. This proves the claim (3.4.13), completing the proof.

$\square$

3.5   Homework Problems for Chapter 3

Problem 1

(a).   Determine whether this function is a linear spline:

$$S(x) = \begin{cases} x, & -1 \leq x \leq 0.5, \\ 0.5 + 2(x - 0.5), & 0.5 \leq x \leq 2, \\ x + 1.5, & 2 \leq x \leq 4. \end{cases}$$

(b).   Do there exist a, b, c and d so that the function

$$S(x) = \begin{cases} ax^3 + x^2 + cx, & -1 \leq x \leq 0, \\ bx^3 + x^2 + dx, & 0 \leq x \leq 1, \end{cases}$$

is a natural cubic spline?

(c).   Determine whether f is a cubic spline with knots $-1, 0, 1$ and 2:

$$f(x) = \begin{cases} 1 + 2(x + 1) + (x + 1)^3, & -1 \leq x \leq 0, \\ 3 + 5x + 3x^2, & 0 \leq x \leq 1, \\ 11 + (x - 1) + 3(x - 1)^2 + (x - 1)^3, & 1 \leq x \leq 2. \end{cases}$$

Problem 2: Constructing a Linear and a Cubic Spline

Given the data set:

t_i	1.2	1.5	1.6	2.0	2.2
y_i	0.4275	1.139	0.8736	−0.9751	−0.1536

a). Let $L(x)$ be the linear spline that interpolates the data. Describe what $L(x)$ consists of, and what conditions it has to satisfy. Find $L(x)$ and compute the value for $L(1.8)$.

b). Let $C(x)$ be the natural cubic spline that interpolates the data. Describe what $C(x)$ consists of, and what conditions it has to satisfy. Find $C(x)$, and compute the value for $C(1.8)$. The computation here can be time consuming, and you may use Matlab to solve the linear system.

Problem 3: Linear Spline in Matlab

Preparation.   Read through the rest of "*A Practical Introduction to Matlab*" by Gockenbach, at the web-page

http://www.math.mtu.edu/~msgocken/intro/intro.pdf

Your task. Write a Matlab function that computes the linear spline interpolation for a given data set. You might need to take a look at the file `cspline_eval.m` in Section 3.4 for some hints. Name your Matlab function `lspline`. This can be defined in the file `lspline.m`, which should begin with:

```
function ls=lspline(t,y,x)
% lspline computes the linear spline
% Inputs:
%      t: vector, contains the knots
%      y: vector, contains the interpolating values at knots
%      x: vector, contains points where the lspline function
%         should be evaluated and plotted
% Output:
%      ls: vector, contains the values of lspline at points x
```

Use your Matlab function `lspline` on the given data set in Problem 2, plot the linear spline for the interval $[1.2, 2.2]$.

What to hand in: Hand in the Matlab file `lspline.m`, and your plots.

Problem 4: Natural Cubic Spline in Matlab

The goal of this problem is to draw Mount Everest with the help of natural cubic splines. We have a rather poor quality photo of the mountain profile, which is given below.

You need to set up a coordinate system, and select a set of knots along the edge of the mountain, and find the coordinates for all these interpolating points. We are aware that the mountain profiles are not very clear in that photo, so please use your imagination when you find these approximate points. You may either print out the mountain picture and work on it on a piece of paper, or send the image to some software and locate the coordinates of the interpolating points. Make sure to select at least 20 points. You may use more points if you see fit.

After you have generated your data set, you need to find a natural cubic spline interpolation. Use the functions `cspline` and `cspline_eval`, which are available in Section 3.4. Read these two functions carefully, try to understand them before using them.

Does it look like there is a smaller peak on the right? Does the peak look rather sharp? How would you deal with this situation, knowing that a single cubic spline function will generate a "smoothest" possible interpolation?

What you need to hand in: A Matlab script that contains your data set, compute the spline functions, and draw the mountain. Also the plot of your Mount Everest.

Have Fun!

Fig. 3.3   Natural cubic spline, for Mount Everest.

Chapter 4

Numerical Integration

4.1 Introduction

In this rather long chapter, we study numerical integration. We focus on single integrals.

Problem Setting: Given a function $f(x)$, defined on an interval $[a, b]$, we want to find an approximate value for the integral

$$I(f) = \int_a^b f(x)\, dx\,.$$

Some motivations and applications include:

- The function f could be very hard to integrate exactly. This happens whenever no explicit expression is available for the antiderivative of f.
- We do not know the exact expression of the function $f(x)$, but we have some algorithm that can compute the values $f(x)$;
- We do not know the function $f(x)$, but we only have a discrete set of data to represent the function.

The main ideas behind all numerical integration methods are as follows:

- We subdivide the large interval $[a, b]$ into many smaller subintervals;
- On each subinterval $[x_i, x_{i+1}]$, we find a polynomial approximation for the function $f(x)$, say $p_i(x) \approx f(x)$ for $x_i \leq x \leq x_{i+1}$;
- We integrate $p_i(x)$ on each subinterval (because polynomials are much easier to integrate), and take the sum over all subintervals. This will provide an approximation for the integral of $f(x)$.

The key step here is to find a suitable polynomial $p_i(x)$ that approximates $f(x)$ on a small subinterval. There are many ways for doing this, using polynomials of various degrees. They lead to many different integration methods.

4.2 Trapezoid Rule

The grid: We partition $[a, b]$ into n subintervals, inserting the points

$$a = x_0 < x_1 < \cdots < x_{n-1} < x_n = b.$$

On each subinterval $[x_i, x_{i+1}]$, we approximate $f(x)$ by a linear polynomial that interpolates the two end points, i.e.,

$$p_i(x_i) = f(x_i), \qquad p_i(x_{i+1}) = f(x_{i+1}).$$

As shown in Figure 4.1, on each subinterval the integral of p_i equals the area of a trapezium:

$$\int_{x_i}^{x_{i+1}} p_i(x)\, dx = \frac{1}{2}(f(x_i) + f(x_{i+1}))(x_{i+1} - x_i).$$

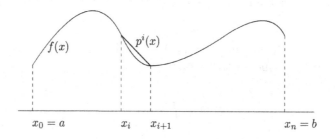

Fig. 4.1   Trapezoid rule: straight line approximation in each subinterval.

Now, we use the approximation

$$\int_{x_i}^{x_{i+1}} f(x)\, dx \approx \int_{x_i}^{x_{i+1}} p_i(x)\, dx$$

$$= \frac{1}{2}\left(f(x_{i+1}) + f(x_i)\right)(x_{i+1} - x_i),$$

and we sum up over all subintervals

$$\int_a^b f(x)\, dx = \sum_{i=0}^{n-1} \int_{x_i}^{x_{i+1}} f(x)\, dx$$

$$\approx \sum_{i=0}^{n-1} \int_{x_i}^{x_{i+1}} p_i(x)\, dx$$

$$= \sum_{i=0}^{n-1} \frac{1}{2}\left(f(x_{i+1}) + f(x_i)\right)(x_{i+1} - x_i).$$

We now consider a *uniform grid* where all subintervals have the same length, and set

$$h = \frac{b-a}{n}, \qquad x_{i+1} - x_i = h.$$

In this case, the previous formula becomes

$$\int_a^b f(x)\,dx \approx \sum_{i=0}^{n-1} \frac{h}{2}\left(f(x_i) + f(x_{i+1})\right)$$

$$= \frac{h}{2}\left[(f(x_0) + f(x_1)) + (f(x_1) + f(x_2)) + \cdots + (f(x_{n-1}) + f(x_n))\right]$$

$$= \frac{h}{2}\left[f(x_0) + 2\sum_{i=1}^{n-1} f(x_i) + f(x_n)\right]$$

$$= h\underbrace{\left[\frac{1}{2}f(x_0) + \sum_{i=1}^{n-1} f(x_i) + \frac{1}{2}f(x_n)\right]}_{T(f;h)}.$$

This gives:

The *trapezoid rule:*

$$\int_a^b f(x) = T(f;h) = h\left[\frac{1}{2}f(x_0) + \sum_{i=1}^{n-1} f(x_i) + \frac{1}{2}f(x_n)\right]. \qquad (4.2.1)$$

Here we used the fact that the boundary points x_0, x_n are counted once, while the inner points x_i with $i = 1, 2, \cdots, n-1$ are counted twice, in the entire summation.

Example 4.1. Let $f(x) = \sqrt{x^2 + 1}$. We want to compute

$$I = \int_{-1}^1 f(x)\,dx$$

using the trapezoid rule. Taking $n = 10$, we set up the data

i	x_i	f_i
0	-1	1.4142136
1	-0.8	1.2806248
2	-0.6	1.1661904
3	-0.4	1.077033
4	-0.2	1.0198039
5	0	1.0
6	0.2	1.0198039
7	0.4	1.077033
8	0.6	1.1661904
9	0.8	1.2806248
10	1	1.4142136

Here $h = 2/10 = 0.2$. By formula (4.2.1), we get

$$T = h\left[\frac{f_0 + f_{10}}{2} + \sum_{i=1}^{9} f_i\right] = 2.3003035.$$

Sample codes. Here are some possible ways to program the trapezoid rule in Matlab. Let a, b, n be given. The function $f(x)$ is also defined: it takes a vector $x = (x_0, x_1, \ldots, x_n)$ and returns the vector $(f(x_0), f(x_1), \ldots, f(x_n))$. For example, $f(x) = x^2 + \sin(x)$ could be defined as:

```
function v=func(x)
  v=x.^2 + sin(x);
end
```

In the following codes, the integral value of trapezoid rule is stored in the variable 'T'. If we use a for-loop for the summation, we can use the following code:

```
h=(b-a)/n;
T = (func(a)+func(b))/2; % the two end-points
for i=1:n-1 % the inner points
  x = a+i*h;
  T = T + func(x);
end
T = T*h;
```

Alternatively, one can use directly the Matlab vector function 'sum', which takes in a vector and adds up all its elements (check out how to use this Matlab function). The code will then be very short and efficient.

```
h=(b-a)/n;
x=[a+h:h:b-h]; % inner points
T = ((func(a)+func(b))/2 + sum(func(x)))*h;
```

Error estimates. We define the error:

$$E_T(f; h) \doteq I(f) - T(f; h)$$

$$= \sum_{i=0}^{n-1} \int_{x_i}^{x_{i+1}} [f(x) - p_i(x)] \, dx$$

$$= \sum_{i=0}^{n-1} E_{T,i}(f; h),$$

where

$$E_{T,i}(f; h) = \int_{x_i}^{x_{i+1}} [f(x) - p_i(x)] \, dx, \qquad (i = 0, 1, \cdots, n-1)$$

is the error on each subinterval.

We know from polynomial interpolation that, for every $x \in [x_i, x_{i+1}]$, one has

$$f(x) - p_i(x) = \frac{1}{2}f''(\xi_i)(x - x_i)(x - x_{i+1}),$$

for some $\xi \in [x_i, x_{i+1}]$ depending on x. Taking the absolute values, we obtain

$$|f(x) - p_i(x)| \leq \frac{1}{2}M_i \cdot (x - x_i)(x_{i+1} - x), \qquad M_i = \max_{\xi \in [x_i, x_{i+i}]} |f''(\xi)|.$$

This gives us an estimate for the error $E_{T,i}(f; h)$ on each subinterval:

$$|E_{T,i}(f; h)| \leq \frac{1}{2}M_i \int_{x_i}^{x_{i+1}} (x - x_i)(x_{i+1} - x)\, dx = \frac{M_i}{12}h^3.$$

(The students are encouraged to work out the details of the integral!)

Setting

$$M \doteq \max_i M_i = \max_{\xi \in [a,b]} |f''(\xi)|,$$

the total error can be estimated as

$$|E_T(f; h)| = \sum_{i=0}^{n-1} |E_{T,i}(f; h)| \leq \sum_{i=0}^{n-1} \frac{M_i}{12}h^3 \leq \frac{h^3}{12}nM = \frac{h^3}{12} \cdot \frac{b-a}{h}M.$$

Recalling the definition of M, we obtain an upper bound on the total error:

$$|E_T(f; h)| \leq \frac{b-a}{12}h^2 \max_{x \in [a,b]} |f''(x)|. \tag{4.2.2}$$

Example 4.2. Consider function $f(x) = e^x$, and the integral

$$I(f) = \int_0^2 e^x\, dx.$$

What is the minimum number of points to be used in the trapezoid rule to ensure an error $\leq 0.5 \times 10^{-4}$?

Answer. We have

$$f'(x) = e^x, \quad f''(x) = e^x, \quad a = 0, \quad b = 2$$

so

$$\max_{x \in (a,b)} |f''(x)| = f''(b) = e^2.$$

By the error bound (4.2.2), it is sufficient to require

$$|E_T(f; h)| \leq \frac{1}{6}h^2 e^2 \leq 0.5 \times 10^{-4}.$$

This gives

$$h^2 \leq 0.5 \times 10^{-4} \times 6 \times e^{-2} \approx 4.06 \times 10^{-5},$$

so

$$\frac{2}{n} = h \leq \sqrt{4.06 \times 10^{-5}} = 0.0064.$$

We get a lower bound on the value of n as

$$n \geq \frac{2}{0.0064} \approx 313.8 \approx 314.$$

To conclude, we need at least $314 + 1 = 315$ points.

4.3   Simpson's Rule

We now explore what happens if we approximate the function f with a second order polynomial, on each subinterval.

We subdivide $[a, b]$ into $2n$ equal subintervals, inserting the points

$$a = x_0 < x_1 < \cdots < x_{2n-1} < x_{2n} = b.$$

As usual, the step size is

$$h = \frac{b - a}{2n}, \qquad x_{i+1} - x_i = h.$$

Note that the number of subintervals here must be an even number, since we will group 2 subintervals together.

Consider the interval $[x_{2i}, x_{2i+2}]$. Notice that this is the union of 2 subintervals. Moreover, it contains the point x_{2i+1} in its interior. We seek a second order polynomial $p_i(x)$ that interpolates $f(x)$ at the three points

$$x_{2i}, \quad x_{2i+1}, \quad x_{2i+2}.$$

See Figure 4.2 for an illustration.

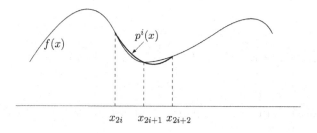

Fig. 4.2   Simpson's rule: quadratic polynomial approximation (thick line) in each subinterval.

Using the Lagrange formula, this quadratic polynomial is computed as

$$p_i(x) = f(x_{2i}) \frac{(x - x_{2i+1})(x - x_{2i+2})}{(x_{2i} - x_{2i+1})(x_{2i} - x_{2i+2})}$$

$$+ f(x_{2i+1}) \frac{(x - x_{2i})(x - x_{2i+2})}{(x_{2i+1} - x_{2i})(x_{2i+1} - x_{2i+2})}$$

$$+ f(x_{2i+2}) \frac{(x - x_{2i})(x - x_{2i+1})}{(x_{2i+2} - x_{2i})(x_{2i+2} - x_{2i+1})}.$$

If the nodes are uniformly spaced, the formula takes the simpler form

$$p_i(x) = \frac{1}{2h^2} f(x_{2i})(x - x_{2i+1})(x - x_{2i+2})$$

$$- \frac{1}{h^2} f(x_{2i+1})(x - x_{2i})(x - x_{2i+2})$$

$$+ \frac{1}{2h^2} f(x_{2i+2})(x - x_{2i})(x - x_{2i+1}).$$

We now compute the integrals (try to fill in the details yourself!)

$$\int_{x_{2i}}^{x_{2i+2}} (x - x_{2i+1})(x - x_{2i+2})\, dx = \frac{2}{3}h^3,$$

$$\int_{x_{2i}}^{x_{2i+2}} (x - x_{2i})(x - x_{2i+2})\, dx = -\frac{4}{3}h^3,$$

$$\int_{x_{2i}}^{x_{2i+2}} (x - x_{2i})(x - x_{2i+1})\, dx = \frac{2}{3}h^3.$$

Using the previous formulas we obtain

$$\int_{x_{2i}}^{x_{2i+2}} p_i(x)\, dx = \frac{1}{2h^2} f(x_{2i}) \int_{x_{2i}}^{x_{2i+2}} (x - x_{2i+1})(x - x_{2i+2})\, dx$$
$$- \frac{1}{h^2} f(x_{2i+1}) \int_{x_{2i}}^{x_{2i+2}} (x - x_{2i})(x - x_{2i+2})\, dx$$
$$+ \frac{1}{2h^2} f(x_{2i+2}) \int_{x_{2i}}^{x_{2i+2}} (x - x_{2i})(x - x_{2i+1})\, dx,$$

and hence

$$\int_{x_{2i}}^{x_{2i+2}} p_i(x)\, dx = \frac{1}{2h^2} f(x_{2i}) \cdot \frac{2}{3}h^3 - \frac{1}{h^2} f(x_{2i+1}) \cdot (-\frac{4}{3}h^3) + \frac{1}{2h^2} f(x_{2i+2}) \cdot \frac{2}{3}h^3$$
$$= \frac{h}{3}\left[f(x_{2i}) + 4f(x_{2i+1}) + f(x_{2i+2}) \right].$$

Summing over all subintervals, we obtain

$$\int_a^b f(x)\, dx \approx S(f; h)$$

$$\doteq \sum_{i=0}^{n-1} \int_{x_{2i}}^{x_{2i+2}} p_i(x)\, dx$$

$$= \frac{h}{3} \sum_{i=0}^{n-1} \left[f(x_{2i}) + 4f(x_{2i+1}) + f(x_{2i+2}) \right].$$

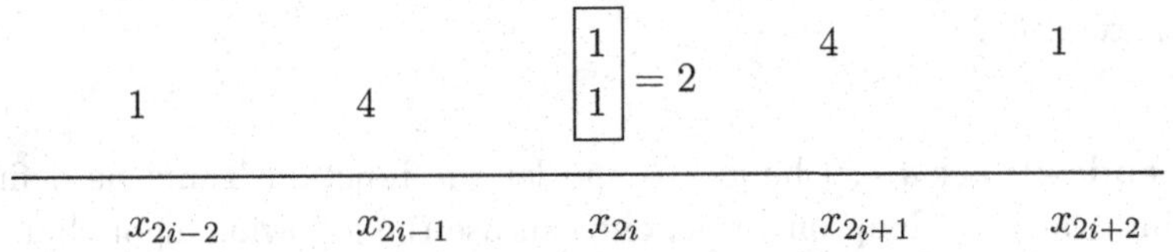

Fig. 4.3 Simpson's rule: adding the constants in each node.

Figure 4.3 shows how to determine the coefficients at each node. We see that for x_0, x_{2n} they are counted once, for odd indices of x_i they are counted 4 times, and for all remaining even indices (that are not at the end points) they are counted twice. In the end, we obtain

---

Simpson's rule:

$$S(f;h) = \frac{h}{3}\left[f(x_0) + 4\sum_{i=1}^{n} f(x_{2i-1}) + 2\sum_{i=1}^{n-1} f(x_{2i}) + f(x_{2n})\right]. \quad (4.3.3)$$

Example 4.3. Let $f(x) = \sqrt{x^2 + 1}$. We want to compute

$$I = \int_{-1}^{1} f(x)\,dx$$

by Simpson's rule. Choosing $n = 5$, we need to evaluate the function at $2n + 1 = 11$ points. We set up the data (same as in Example 4.1):

i	x_i	f_i
0	-1	1.4142136
1	-0.8	1.2806248
2	-0.6	1.1661904
3	-0.4	1.077033
4	-0.2	1.0198039
5	0	1.0
6	0.2	1.0198039
7	0.4	1.077033
8	0.6	1.1661904
9	0.8	1.2806248
10	1	1.4142136

Here $h = 2/10 = 0.2$. Using Simpson's formula we obtain

$$S(f;0.2) = \frac{h}{3}\left[f_0 + 4(f_1 + f_3 + f_5 + f_7 + f_9) + 2(f_2 + f_4 + f_6 + f_8) + f_{10}\right]$$
$$= 2.2955778.$$

This is somewhat smaller than the number we get with trapezoid rule, and it is actually more accurate. Can you intuitively explain why this happens, for this particular example?

Sample codes: Let a, b, n be given, and let the function 'func' be defined. To find the integral with Simpson's rule, one can use the following algorithm:

```
h=(b-a)/2/n;
xodd=[a+h:2*h:b-h]; % x_i with odd indices
xeven=[a+2*h:2*h:b-2*h]; % x_i with even indices, inner points
SI=(h/3)*(func(a)+4*sum(func(xodd))+2*sum(func(xeven))+func(b));
```

Error estimate. One can prove that the error on each subinterval $[x_{2i}, x_{2i+2}]$ is bounded by

$$|E_{S,i}(f;h)| = \left| \int_{x_{2i}}^{x_{2i+2}} [f(x) - p_i(x)]\,dx \right| \leq \frac{h^5}{90} M_i, \qquad (4.3.4)$$

where

$$M_i = \max_{\xi \in [x_{2i}, x_{2i+2}]} \left| f^{(4)}(\xi) \right|.$$

Summing over all subintervals and setting

$$M = \max_i M_i = \max_{\xi \in [a,b]} |f^{(4)}(\xi)|,$$

we can bound the total error as

$$|E_S(f;h)| = |I(f) - S(f;h)|$$

$$\leq \frac{h^5}{90} \sum_{i=0}^{n-1} M_i$$

$$\leq \frac{h^5}{90} nM$$

$$= \frac{h^5}{90} \cdot \frac{b-a}{2h} M.$$

This gives us the error bound

$$|E_S(f;h)| \leq \frac{b-a}{180} h^4 \max_{\xi \in [a,b]} \left| f^{(4)}(\xi) \right|. \qquad (4.3.5)$$

Proof. (for (4.3.4), optional) (This proof is only a formal proof, not rigorous. A more accurate proof could be derived using the error form of polynomial interpolation, in Chapter 2.) For notational simplicity, let us consider an interval $[a, a+2h]$, and approximate the integral by

$$\int_a^{a+2h} f(x)\,dx \approx \frac{h}{3}\left[f(a) + 4f(a+h) + f(a+2h) \right].$$

Using the Taylor expansions for f:

$$f(a+h) = f(a) + hf'(a) + \frac{h^2}{2} f''(a) + \frac{h^3}{6} f'''(a) + \frac{h^4}{24} f^{(4)}(a) + \cdots$$

$$f(a+2h) = f(a) + 2hf'(a) + 2h^2 f''(a) + \frac{4h^3}{3} f'''(a) + \frac{2h^4}{3} f^{(4)}(a) + \cdots,$$

we get

$$f(a) + 4f(a+h) + f(a+2h) = 6f(a) + 6hf'(a) + 4h^2 f''(a) + 2h^3 f'''(a) + \frac{5}{6} h^4 f^{(4)}(a) + \cdots,$$

therefore

$$\frac{h}{3}\left[f(a) + 4f(a+h) + f(a+2h) \right]$$

$$= 2hf(a) + 2h^2 f'(a) + \frac{4}{3} h^3 f''(a) + \frac{2}{3} h^4 f'''(a) + \frac{5}{18} h^5 f^{(4)}(a) + \cdots.$$

Now we go back to the original integral. We observe that

$$\int_a^{a+2h} f(x)\,dx = \int_0^{2h} f(a+s)\,ds.$$

Using the Taylor expansion

$$f(a+s) = f(a) + sf'(a) + \frac{s^2}{2}f''(a) + \frac{s^3}{6}f'''(a) + \frac{s^4}{24}f^{(4)}(a) + \cdots$$

and computing the integral of each term, we obtain

$$\int_0^{2h} f(a+s)\,ds$$

$$= \int_0^{2h} \left[f(a) + sf'(a) + \frac{s^2}{2}f''(a) + \frac{s^3}{6}f'''(a) + \frac{s^4}{24}f^{(4)}(a) + \cdots \right] ds$$

$$= 2hf(a) + f'(a)\int_0^{2h} s\,ds + f''(a)\int_0^{2h} \frac{s^2}{2}\,ds + f'''(a)\int_0^{2h} \frac{s^3}{6}\,ds$$

$$+ f^{(4)}(a)\int_0^{2h} \frac{s^4}{24}\,ds + \cdots$$

$$= 2hf(a) + 2h^2 f'(a) + \frac{4}{3}h^3 f''(a) + \frac{2}{3}h^4 f'''(a) + \frac{4}{15}h^5 f^{(4)}(a) + \cdots.$$

Comparing this with the Simpson's rule, we get the estimate

$$E_{S,i} = \left[\frac{4}{15} - \frac{5}{18} \right] h^5 f^{(4)}(a) + \cdots$$

$$= -\frac{1}{90} h^5 f^{(4)}(a) + \cdots$$

$$= -\frac{1}{90} h^5 f^{(4)}(\xi),$$

for some $\xi \in [a, a+2h]$. Taking absolute values, one obtains (4.3.4). □

Example 4.4. We want to evaluate the integral of $f(x) = e^x$ on the interval $[0,2]$, using Simpson's rule. In order to achieve an error $\leq 0.5 \times 10^{-4}$, how many points must we take?

Answer. Using the error bound (4.3.5), we have

$$|E_S(f;h)| \leq \frac{2}{180} h^4 e^2 \leq 0.5 \times 10^{-4}$$

$$\Rightarrow h^4 \leq 45/e^2 \times 10^{-4} = 6.09 \times 10^{-4}$$

$$\Rightarrow h \leq 0.1571$$

$$\Rightarrow n = \frac{b-a}{2h} = 6.36 \approx 7.$$

In conclusion, we need at least $2n+1 = 15$ points.

Recall that, using the trapezoid rule, to achieve the same accuracy we needed at least 314 points. Simpson's rule uses much fewer points. Here we see the power of a higher order method!

4.4   Recursive Trapezoid Rule

Here we design various levels of the same rule, using subintervals of different sizes. The rules with higher number of subintervals are built on top of those with fewer number of subintervals. Such methods are also called *composite schemes.* Although such composite schemes could be derived for many basic rules, we will use trapezoid rule to introduce the concept.

Divide $[a, b]$ into 2^m equal subintervals, for various values of m, see Figure 4.4:

$$h_m = \frac{b-a}{2^m}, \qquad h_{m+1} = \frac{1}{2}h_m.$$

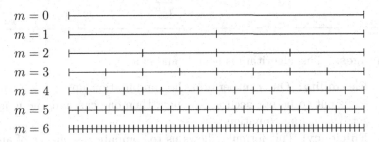

Fig. 4.4   Recursive division of intervals, first few levels. Each new level is obtained by cutting each subinterval further into two subintervals.

At level m and $m + 1$, we have

$$T(f; h_m) = h_m \cdot \left[\frac{1}{2}f(a) + \frac{1}{2}f(b) + \sum_{i=1}^{2^m-1} f(a + ih_m)\right],$$

$$T(f; h_{m+1}) = h_{m+1} \cdot \left[\frac{1}{2}f(a) + \frac{1}{2}f(b) + \sum_{i=1}^{2^{m+1}-1} f(a + ih_{m+1})\right].$$

Recall that, in any algorithm, the computation that is most costly is the function evaluation $f(x)$, because the computer computes most functions using a truncated Taylor series, taking the sum of a large number of terms. For this reason, to design an efficient algorithm, one tries to minimize the number of function evaluations.

Assume that $T(f; h_m)$ is already computed, and we want to compute $T(f; h_{m+1})$. Observe that in the computation of $T(f; h_{m+1})$, about half of the function evaluations $f(x_i)$ were already done in the computation of $T(f; h_m)$. To save computation time, it is reasonable to use $T(f; h_m)$ again in some way, and only perform the function evaluation at the new points in $T(f; h_{m+1})$.

We can rearrange the terms in $T(f; h_{m+1})$. By an algebraic manipulation we get

$$T(f; h_{m+1})$$

$$= \frac{h_m}{2} \left[\frac{1}{2} f(a) + \frac{1}{2} f(b) + \sum_{i=1}^{2^m - 1} f(a + ih_m) + \sum_{j=0}^{2^m - 1} f(a + (2j+1)h_{m+1}) \right]$$

$$= \frac{1}{2} T(f; h_m) + h_{m+1} \sum_{j=0}^{2^m - 1} f(a + (2j+1)h_{m+1}).$$

This leads to the *Recursive Trapezoid Rule*:

$$T(f; h_{m+1}) = \frac{1}{2} T(f; h_m) + h_{m+1} \sum_{j=0}^{2^m - 1} f(a + (2j+1)h_{m+1}). \qquad (4.4.6)$$

Advantages.   This algorithm is flexible and efficient.

1. **Flexibility:** One can perform the computation up to a level m. If this turns out to be not accurate enough, then one can add one more level to get better approximation.
2. **Efficiency:** This formula allows us to compute a sequence of approximations to a definite integral using the trapezoid rule without re-evaluating the integrand at points where it has already been evaluated.

A very interesting aspect, discussed in the next section, is that this sequence of approximations allows a further development of algorithm, using Richardson's extrapolation technique. This leads to the Romberg algorithm, providing approximations of high order.

4.5   Romberg Algorithm

Assume that we have generated a sequence of approximations, with different values of h, using the Composite Trapezoid Rule described in the previous section. We denote these values as

$$T(f; h), \quad T(f; h/2), \quad T(f; h/4), \quad \cdots$$

We now explore how one could combine these numbers in particular ways, and get higher order approximations to the exact integral $I(f)$.

The particular form of the final algorithm will depend on the error formula. Consider the trapezoid rule. If f admits a Taylor series expansion, then one can prove that the error satisfies the Euler-MacLaurin's formula

$$E(f; h) = I(f) - T(f; h) = a_2 h^2 + a_4 h^4 + a_6 h^6 + \cdots .$$

Here a_n depends on the derivatives $f^{(n)}$. Note that only even powers of h show up in the error formula! Due to symmetry of the methods, all terms containing h^k, with k odd, are cancelled out!

When we reduce the grid size h by a half, the error formula becomes

$$E\left(f; \frac{h}{2}\right) = I(f) - T\left(f; \frac{h}{2}\right)$$

$$= a_2 \left(\frac{h}{2}\right)^2 + a_4 \left(\frac{h}{2}\right)^4 + a_6 \left(\frac{h}{2}\right)^6 + \cdots .$$

We now have

$$I(f) = T(f; h) + a_2 h^2 + a_4 h^4 + a_6 h^6 + \cdots , \qquad (4.5.7)$$

$$I(f) = T\left(f; \frac{h}{2}\right) + a_2 \frac{h^2}{2^2} + a_4 \frac{h^4}{2^4} + a_6 \frac{h^6}{2^6} + \cdots . \qquad (4.5.8)$$

Our goal is to combine the two approximations $T(f; h)$ and $T(f; \frac{h}{2})$ to get new approximation which is more accurate. In essence, we wish to cancel the leading error terms, i.e., the ones with h^2 in (4.5.7)-(4.5.8). This can be achieved by a simple algebraic manipulation.

Multiplying (4.5.8) by $2^2 = 4$ and subtracting (4.5.7), we get

$$3 \cdot I(f) = 4 \cdot T(f; h/2) - T(f; h) + a_4' h^4 + a_6' h^6 + \cdots ,$$

then

$$I(f) = \underbrace{\frac{4}{3} T(f; h/2) - \frac{1}{3} T(f; h)}_{U(h)} + \tilde{a}_4 h^4 + \tilde{a}_6 h^6 + \cdots .$$

Here a_k' and $\tilde{a}_k$ are coefficients only depending on f^k, not on h.

Note that $U(h)$ is of 4th order accuracy because we cancelled out the second order error terms! This is much better than $T(f; h)$, which is only second order. We now write:

$$U(h) = \frac{2^2 T(f; h/2) - T(f; h)}{2^2 - 1}. \qquad (4.5.9)$$

This procedure is called the *Richardson extrapolation*.

The extrapolation need not stop here. It can be iterated, achieving even higher accuracy. Using (4.5.9) for different values of h, we can compute $U(h/2), U(h/4), U(h/8), \cdots$.

The same extrapolation idea can be applied again to these values $U(h/2), U(h/4), U(h/8), \ldots$. We have

$$I(f) = U(h) + \tilde{a}_4 h^4 + \tilde{a}_6 h^6 + \cdots , \qquad (4.5.10)$$

$$I(f) = U(h/2) + \tilde{a}_4 (h/2)^4 + \tilde{a}_6 (h/2)^6 + \cdots . \qquad (4.5.11)$$

To cancel the term containing the factor h^4, we multiply (4.5.11) by 2^4 and subtract (4.5.10). This yields

$$(2^4 - 1)I(f) = 2^4 U(h/2) - U(h) + \tilde{a}_6' h^6 + \cdots .$$

Define

$$V(h) = \frac{2^4 U(h/2) - U(h)}{2^4 - 1}.$$

Then

$$I(f) = V(h) + \tilde{a}_6' h^6 + \cdots.$$

So $V(h)$ yields an approximation of order 6, which is even better than $U(h)$.

One can repeat this procedure several times, which leads to the following.

Romberg algorithm. Setting $H = b - a$, we initialize the algorithm by computing the first column of data as follows:

$$
\begin{aligned}
R(0,0) &= T(f;H) = \frac{H}{2}(f(a) + f(b)), \\
R(1,0) &= T(f;2^{-1}H), \\
R(2,0) &= T(f;2^{-2}H), \\
&\ \ \vdots \\
R(m,0) &= T(f;2^{-m}H).
\end{aligned}
$$

Here $R(i,0)$ for $i = 0, 1, \cdots, m$ could be computed by the recursive trapezoid formula. Notice that this initial column (with $k = 0$) has $(m + 1)$ entries. Based on this initial column, we now compute the following columns for $k = 1, 2, 3, \ldots$, by an inductive procedure.

Assume that column $k - 1$ has been constructed. Then the entry number j in column k, denoted by $R(j, k)$, is computed by the formula

$$R(j,k) = R(j,k-1) + \frac{R(j,k-1) - R(j-1,k-1)}{2^{2k} - 1}. \qquad (4.5.12)$$

The algorithm can be carried out either column-by-column or row-by-row. We see that column $k = 1$ has m entries, column $k = 2$ has $m - 1$ entries, and so on and so forth. Finally, column $k = m$ will have 1 entry, and the algorithm stops. This generates a set of numbers having triangular shape, which is called the *Romberg triangle*. See Figure 4.5 for an illustration.

By the extrapolation analysis, the accuracy of the term $R(j, k)$ is formally

$$I(f) = R(j,k) + \mathcal{O}(h^{2(k+1)}), \qquad h = \frac{H}{2^k}. \qquad (4.5.13)$$

Here we give a pseudo-code, using a column-by-column algorithm.

$R =$ romberg(f, a, b, n)

$R = n \times n$ matrix

$h = b - a; \ R(1,1) = [f(a) + f(b)] * h/2;$

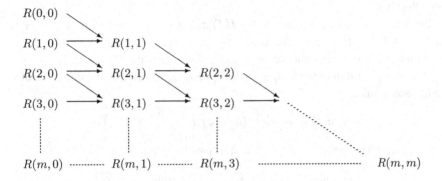

Fig. 4.5 Romberg triangle

for $i = 1$ to $n - 1$ do   %1st column recursive trapezoid
$\quad R(i + 1, 1) = R(i, 1)/2$;
$\quad h = h/2$;
$\quad$for $k = 1$ to $2^i - 1$ step 2 do
$\quad\quad R(i + 1, 1) = R(i + 1, 1) + h * f(a + kh)$
$\quad$end

end
for $j = 2$ to n do   %2 to n column
$\quad$for $i = j$ to n do
$\quad\quad R(i, j) = R(i, j - 1) + \frac{1}{4^{j-1}-1}[R(i, j - 1) - R(i - 1, j - 1)]$
$\quad$end

end

4.6 Adaptive Simpson's Quadrature Scheme

In a numerical integration, we expect that the error will be large in regions where the function varies a lot, i.e., where the derivatives $f^{(k)}$ are big. Using a uniform grid, the error may not be evenly distributed, and we have little control over the accuracy. For this reason, it is convenient to insert more points in regions where the derivatives $f^{(k)}$ are large.

In this section, we describe an algorithm that will automatically decide how many points should be inserted, in order to achieve the required error tolerance. Such an algorithm is called an *adaptive scheme*, and the corresponding grid is an *adaptive grid*. In any adaptive algorithm, the key step is an error estimate.

We illustrate this idea using Simpson's rule. (The same idea could also be adapted to the trapezoid rule. But since Simpson's rule is higher order, it is far more effective.)

Given an interval $[a, b]$, denote by $I(f)[a, b]$ the exact integral of f over $[a, b]$, and let $\varepsilon > 0$ be the error tolerance. More precisely, we require that the error in the numerical integration should be smaller than $\varepsilon(b - a)$.

Let $h = \frac{b-a}{2}$ and denote by $S_1[a, b]$ the approximate value of the integral obtained by Simpson's rule:

$$S_1[a, b] = \frac{b - a}{6}\left[f(a) + 4f\left(\frac{a + b}{2}\right) + f(b)\right].$$

Recall that the error for this basic step can be written as

$$E_1[a, b] = -\frac{1}{90}h^5 f^{(4)}(\xi), \qquad \text{for some} \quad \xi \in [a, b].$$

We can write

$$I(f)[a, b] = S_1[a, b] + E_1[a, b].$$

We now split $[a, b]$ in two equal subintervals, inserting the mid-point $c = \frac{a+b}{2}$. We have

$$
\begin{aligned}
I(f)[a, b] &= I(f)[a, c] + I(f)[c, b] \\
&= S_1[a, c] + E_1[a, c] + S_1[c, b] + E_1[c, b] \\
&= S_2[a, b] + E_2[a, b],
\end{aligned}
$$

where

$$S_2[a, b] \doteq S_1[a, c] + S_1[c, b]$$

and

$$
\begin{aligned}
E_2[a, b] &\doteq E_1[a, c] + E_1[c, b] \\
&= -\frac{1}{90}(h/2)^5\left[f^{(4)}(\xi_1) + f^{(4)}(\xi_2)\right],
\end{aligned}
$$

for some $\xi_1 \in [a, c]$ and $\xi_2 \in [c, b]$.

Note that the values $S_1[a, c], S_1[c, b], S_2[a, b]$ can be computed, but the values $E_1[a, c], E_1[c, b], E_2[a, b]$ are unknown. Our goal is to get some good estimates on these errors, using the computed values of $S_1[a, c], S_1[c, b], S_2[a, b]$.

Assume $f^{(4)}$ does NOT change much over the small interval $[a, b]$, i.e.,

$$f^{(4)}(\xi) \approx f^{(4)}(\xi_1) \approx f^{(4)}(\xi_2) \qquad \text{for all} \quad \xi \in [a, b].$$

Then $E_1[a, c] \approx E_1[c, b]$, and

$$E_2[a, b] \approx 2E_1[a, c] = 2\frac{1}{2^5}E_1[a, b] = \frac{1}{2^4}E_1[a, b].$$

This gives

$$
\begin{aligned}
S_2[a, b] - S_1[a, b] &= (I(f) - E_2[a, b]) - (I(f) - E_1[a, b]) \\
&= E_1 - E_2 \\
&\approx 2^4 E_2 - E_2 \\
&= 15E_2.
\end{aligned}
$$

This means, we can estimate the error E_2 using S_2 and S_1, as

$$E_2 \approx \frac{1}{15}(S_2 - S_1).$$

In order to have $|E_2| \le \varepsilon(b - a)$, we need

$$\frac{|S_2 - S_1|}{2^4 - 1} \le \varepsilon(b - a).$$

This gives us an estimate of the error on every subinterval, allowing us to design an *adaptive recursive* algorithm. If the local error on a certain subinterval shall be too big, we reject the computation, cut the interval into 2 smaller subinterval (i.e., make a finer grid), and repeat the procedure.

Once the error tolerance is met on a subinterval, one can manipulate further and get an even better result. We have

$$I(f) = S_1 + E_1,$$
$$I(f) = S_2 + E_2 \approx S_2 + \frac{1}{2^4}E_1.$$

Multiplying the second equation by 2^4 and subtracting it from the first, we obtain

$$I(f) \approx \frac{2^4 S_2 - S_1}{2^4 - 1} = S_2 + \frac{S_2 - S_1}{15}.$$

Note that this approximation gives the best approximation when $f^{(4)}$ is a constant.

Pseudocode: Let f denote the function, $[a, b]$ the interval, and ε the total tolerance for error. Below is a pseudo-code.

```
answer=Simpson(f, a, b, ε)
compute S₁ and S₂
If |S₂ - S₁| < 15ε(b - a)
        answer= S₂ + (S₂ - S₁)/15;
else
        c = (a + b)/2;
        Lans=Simpson(f, a, c, ε/2);
        Rans=Simpson(f, c, b, ε/2);
        answer=Lans+Rans;
end
```

Note that the program "Simpson" calls itself. Such a feature could not be implemented directly in Matlab. It could be coded in Fortran, C, or C++, and then one can link the complied file into Matlab.

In Matlab, one can use **quad** to compute numerical integration, using an adaptive algorithm. Try **help quad**, it will give you information on it. One can call the program by using:

$$a=\texttt{quad('fun',a,b,tol)}$$

It uses adaptive Simpson's formula.

See also `quad8`, another higher order method.

Remark: One could also design an adaptive trapezoid method. Would you like to try it and work out the algorithm?

Of course, due to its higher accuracy, Simpson's rule is the most commonly used one, even in an adaptive form.

4.7   Gaussian Quadrature Formulas

All the numerical integration rules we have learned so far have the form

$$\int_a^b f(x)\,dx \approx R(f) \doteq A_0 f(x_0) + A_1 f(x_1) + \cdots + A_n f(x_n). \qquad (4.7.14)$$

We use $R(f)$ to represent the rule. Here $x_i \in [a,b]$ are called the *nodes*, and A_i's are the weights.

For example, the trapezoid rule is:

$$x_0 = a, \quad x_1 = b, \quad A_0 = A_1 = \frac{b-a}{2},$$

and the Simpson's rule corresponds to

$$x_0 = a, \quad x_1 = \frac{a+b}{2}, \quad x_2 = b, \quad A_0 = A_2 = (b-a)\frac{1}{6}, \quad A_1 = (b-a)\frac{2}{3}.$$

Recall the derivation of these rules: we first fix the nodes x_i, then we adjust the weights A_i to achieve the highest possible accuracy.

New idea: If now we allow the nodes x_i also to be adjusted, we gain more freedom, and may obtain higher accuracy as well. Below we explain the details.

Gaussian quadrature: Fix an integer $n \geq 0$. We want to find $n+1$ nodes x_i and weights A_i ($i = 0, 1, 2, \cdots, n$) so that the formula (4.7.14) gives the exact value of the integral for polynomial functions $f(x)$ of highest possible degree m. In other words,

$$\int_a^b f(x)\,dx = R(f) \doteq A_0 f(x_0) + A_1 f(x_1) + \cdots + A_n f(x_n), \qquad \text{if } f \in P^m. \quad (4.7.15)$$

Here, m is called the *degree of precision*. In the end, m will be exactly the order of the method.

A polynomial of degree m can be written as

$$p_m(x) = a_m x^m + a_{m-1} x^{m-1} + \cdots + a_1 x + a_0.$$

By the linearity of the integration operator, we have

$$\int_a^b p_m(x)\, dx = a_m \int_a^b x^m\, dx + a_{m-1} \int_a^b x^{m-1}\, dx + \cdots + a_1 \int_a^b x\, dx + a_0 \int_a^b 1\, dx.$$

Since the integration rule $R(f)$ is also a linear operator, we have

$$R(p_m) = a_m R(x^m) + a_{m-1} R(x^{m-1}) + \cdots + a_1 R(x) + a_0 R(1).$$

If the rule is exact for all functions x^k, with $k = 0, 1, \cdots, m$, i.e.,

$$R(x^k) = \int_a^b x^k\, dx, \qquad k = 0, 1, \cdots, m, \tag{4.7.16}$$

then

$$\begin{aligned}
R(p_m) &= a_m \int_a^b x^m\, dx + a_{m-1} \int_a^b x^{m-1}\, dx + \cdots + a_1 \int_a^b x\, dx + a_0 \int_a^b 1\, dx \\
&= \int_a^b p_m(x)\, dx.
\end{aligned}$$

This shows that the rule is also exact for any polynomial of degree m.

In order to verify the identity (4.7.15) for all polynomials of degree m, it suffices to check that it holds for all monomials $f(x) = x^k$, with $k = 0, 1, \cdots, m$. These are the basis functions for polynomials. This is a very useful observation, which simplifies significantly our discussion.

In our problem, we need to determine $2(n + 1)$ unknowns: $x_0, \ldots, x_n$ and $A_0, \ldots, A_n$, while the number of equations in (4.7.16) is $m + 1$. To achieve a unique solution we thus need $2(n + 1) = m + 1$, hence $m = 2n + 1$.

To fix the ideas, we consider the interval $[-1, 1]$. In the end we will extend the algorithm to a general interval of $[a, b]$.

Let us start with $n = 1$. In this case we must determine 2 nodes x_0, x_1 and 2 weights A_0, A_1. Our rule takes the form

$$\int_{-1}^1 f(x)\, dx = R_1(f) = A_0 f(x_0) + A_1 f(x_1). \tag{4.7.17}$$

If (4.7.17) is exact for polynomials of degree m, where $m = 2 + 1 = 3$, then it must be exact for functions $1, x, x^2, x^3$. This gives us 4 equations. Recall that

$$\int_{-1}^1 x^k\, dx = \begin{cases} \dfrac{2}{k+1}, & k \text{ even,} \\[2mm] 0, & k \text{ odd.} \end{cases}$$

This leads to

$$\begin{aligned}
f(x) = 1: \quad & A_0 + A_1 = 2, & (4.7.18) \\
f(x) = x: \quad & A_0 x_0 + A_1 x_1 = 0, & (4.7.19) \\
f(x) = x^2: \quad & A_0 x_0^2 + A_1 x_1^2 = \frac{2}{3}, & (4.7.20) \\
f(x) = x^3: \quad & A_0 x_0^3 + A_1 x_1^3 = 0. & (4.7.21)
\end{aligned}$$

These are nonlinear equations, and in general nonlinear equations are not easy to solve. We first observe some symmetry properties. Writing (4.7.19) and (4.7.21) as

$$A_0 x_0 = -A_1 x_1, \qquad A_0 x_0^3 = -A_1 x_1^3,$$

we obtain

$$x_0^2 = x_1^2, \qquad \text{hence} \qquad x_0 = -x_1.$$

This means that the two nodes are positioned symmetrically around the origin. Plugging this back into (4.7.19), we obtain

$$A_0 x_0 - A_1 x_0 = 0, \qquad \text{hence} \qquad A_0 = A_1,$$

indicating that the weights are also symmetrical about the origin.

It is now easy to solve this system, and obtain

$$x_0 = -\frac{1}{\sqrt{3}}, \quad x_1 = \frac{1}{\sqrt{3}}, \quad A_0 = 1, \quad A_1 = 1.$$

This gives us the Gaussian quadrature rule for $N = 1$:

$$\int_{-1}^{1} f(x)\, dx \approx R_1(f) = f\left(-\frac{1}{\sqrt{3}}\right) + f\left(\frac{1}{\sqrt{3}}\right). \tag{4.7.22}$$

This will have degree of precision $m = 3$.

Next, let us consider the case $n = 2$. We need to find nodes x_0, x_1, x_2 and weights A_0, A_1, A_2, in such a way that the rule

$$\int_{-1}^{1} f(x)\, dx = R_2(f) = A_0 f(x_0) + A_1 f(x_1) + A_2 f(x_2), \tag{4.7.23}$$

gives the exact integral for all polynomials of degree $m = 2n + 1 = 5$.

By symmetry considerations, we deduce

$$x_0 = -x_2, \quad x_1 = 0, \quad A_0 = A_2.$$

Hence the rule (4.7.23) must have the form

$$\int_{-1}^{1} f(x)\, dx = R_2(f) = A_0 f(x_0) + A_1 f(0) + A_0 f(-x_0).$$

We observe that the above rule already gives the correct integral (i.e., zero) whenever $f(x) = x^k$ with k odd. Indeed,

$$R_2(x^k) = A_0 x_0^k + A_1 0^k - A_0 x_0^k = 0 = \int_{-1}^{1} x^k\, dx.$$

It remains to determine A_0, A_1, x_0, requiring that the rule gives the exact integral for $f(x) = 1, x^2, x^4$:

$$f(x) = 1: \quad 2A_0 + A_1 = 2,$$

$$f(x) = x^2: \quad 2A_0 x_0^2 = \frac{2}{3},$$

$$f(x) = x^4: \quad 2A_0 x_0^4 = \frac{2}{5}.$$

Table 4.1 Gaussian Quadrature Nodes and Weights

n	Nodes x_i	Weights A_i
1	0	2
2	$\pm\sqrt{1/3}$	1
3	$\pm\sqrt{0.6}$	$5/9$
	0	$8/9$
4	$\pm\sqrt{(3-\sqrt{4.8})/7}$	$(18+\sqrt{30})/36$
	$\pm\sqrt{(3+\sqrt{4.8})/7}$	$(18-\sqrt{30})/36$
5	$\pm\sqrt{(5-\sqrt{40/7})/9}$	$(322+13\sqrt{70})/900$
	0	$128/225$
	$\pm\sqrt{(5+\sqrt{40/7})/9}$	$(322-13\sqrt{70})/900$

These three equations can now be easily solved, obtaining

$$x_0 = -\sqrt{3/5}, \qquad x_1 = 0, \qquad x_2 = \sqrt{3/5},$$

and

$$A_0 = A_2 = 5/9, \qquad A_1 = 8/9.$$

This yields the Gaussian quadrature rule for $N = 2$:

$$\int_{-1}^{1} f(x)\,dx \approx R_2(f) = \frac{1}{9}\left[5f\left(-\sqrt{3/5}\right) + 8f(0) + 5f\left(\sqrt{3/5}\right)\right]. \qquad (4.7.24)$$

This rule will have degree of precision $m = 5$.

One can prove that, for every $n \geq 0$, the corresponding nodes $x_0, x_1, \ldots, x_n$ are precisely the roots of the *Legendre polynomials* $q_{n+1}(x)$. On the interval $[-1, 1]$, the first four Legendre polynomials are:

$$q_0(x) = 1, \qquad q_1(x) = x, \qquad q_2(x) = \frac{3}{2}x^2 - \frac{1}{2}, \qquad q_3(x) = \frac{5}{2}x^3 - \frac{3}{2}x.$$

Their roots are

$$q_1 : 0, \qquad q_2 : \pm 1/\sqrt{3}, \qquad q_3 : 0, \quad \pm\sqrt{3/5}.$$

After the points $x_0 < x_1 < \cdots x_n$ have been determined, the weights A_i can be computed as

$$A_i = \int_{-1}^{1} l_i(x)\,dx, \qquad l_i(x) = \prod_{j\neq i} \frac{x - x_j}{x_i - x_j}.$$

Here $l_i(x)$ is the *cardinal function* for the point x_i.

See Table 4.1 for the list of all Gaussian quadrature nodes and weights for $n \leq 5$.

If instead of the interval $[-1, 1]$ we have a general interval $[a, b]$, we can use the linear transformation

$$x = \frac{2t - (a+b)}{b-a}, \qquad t = \frac{1}{2}(b-a)x + \frac{1}{2}(a+b).$$

Notice that, as $x \in [-1, 1]$, we have $t \in [a, b]$.

Each node x_i with weight A_i on the interval $[-1, 1]$ (see Table 4.1) is then transformed into a new node t_i with weight $\bar{A}_i$ on the interval $[a, b]$, by the formulas

$$t_i = \frac{1}{2}(b-a)x_i + \frac{1}{2}(a+b), \qquad \bar{A}_i = \frac{b-a}{2}A_i.$$

Advantages:

(1). Higher order accuracy with minimum number of nodes.

(2). Since all nodes are in the interior of the interval, these formulas can handle integrals of functions that approach infinity at one end of the interval (provided that the integral is defined). For example:

$$\int_0^1 \frac{1}{\sqrt{x}}\, dx.$$

4.8   Matlab Simulations

We now talk about effective programming in Matlab. We want to compute the vector product of two vectors:

$$z_i = x_i \cdot y_i, \qquad i = 1, 2, \ldots, n.$$

We use three different programs to achieve this same goal.

Method 1:   We do not allocate memory space for z in advance, and we compute the elements in the new vector by a for-loop.

```
x = rand(n,1);     % random vector
y = rand(n,1);     % of length n
clear z;
t = cputime;
for i=1:n
    z(i) = x(i)*y(i);
end
cputime-t
```

Method 2:   We allocate space for z in advance, but we still use a for loop.

```
z = zeros(n,1);
for i=1:n
    z(i) = x(i)*y(i);
end
```

Method 3:   We use vector operation in Matlab directly.

```
z = x.*y;
```

The CPU-times measured in seconds for these three methods are listed in Table 4.2.

Table 4.2   CPU-times for three methods.

n	Method 1	Method 2	Method 3
5000	2.86	0.24	0.00
10000	14.22	0.49	0.00
20000	59.65	0.97	0.01
100000	–	4.87	0.03
1000000	–	48.84	0.30

The moral of this story: Use pre-defined Matlab functions if possible!

Romberg integration: numerical simulations   We perform numerical integration for

$$\int_0^{\pi/2} \cos(2x)e^{-x}dx.$$

The exact value for the integral is about 0.2415759.

The Romberg algorithm with Trapezoid as the base rule gives the following result for the approximations:

```
0.6221
0.3111   0.2074
0.2575   0.2397   0.2419
0.2455   0.2415   0.2416   0.2416
```

The errors are:

```
3.80e-01
6.94e-02   3.41e-02
1.59e-02   1.87e-03   2.81e-04
3.90e-03   1.11e-04   5.85e-06   1.47e-06
```

If we expand the integration interval to 2π, the exact value is about 0.1996265.... The Romberg triangle looks like:

```
3.1475
1.7095   1.2302
0.5141   0.1156   0.0413
0.2570   0.1714   0.1751   0.1772
```

The errors are:

```
2.94e+00
1.50e+00    1.03e+00
3.14e-01    8.39e-02    1.58e-01
5.74e-02    2.82e-02    2.45e-02    2.24e-02
```

Matlab's adaptive Simpson's method: quad can be called with the following syntax:

$\qquad$ q = quad('f',a,b,tolerance,trace)

or

$\qquad$ q = quad(@f,a,b,tolerance,trace)

$\qquad$ If you set the option *trace* $\neq 0$, Matlab will show the development of the integration.

```
quad('f3',0,2*pi,1.e-6,1)
```

or

```
quad(@f3,0,2*pi,1.e-6,1)
```

Matlab's recursive numerical integration are quadl and quad8. The syntax and usage are the same as for **quad**. The functions use a high order recursive algorithm. Usually, **quadl** is preferred over **quad8**.

4.9   A More Abstract Discussion (optional)

We now discuss these rules in a more abstract setting. Consider a quadrature rule for integrating $f(x)$ over the interval $[-1, 1]$. In essence, all these rules are based on polynomial interpolation through the nodes. Let $x_0, x_1, \cdots, x_n$ denote the nodes.

Closed Newton-Cotes Formulas.   We first consider a set of uniformly spaced nodes which includes the two endpoints $-1, 1$, so that

$$h = \frac{2}{n}, \qquad x_i = -1 + ih, \qquad i = 0, 1, 2, \cdots, n. \qquad (4.9.25)$$

The rules derived using these points are called *closed Newton-Cotes Formulas*. The polynomial that interpolates $f(x)$ through these nodes in Lagrange form is

$$p_n(x) = f(x_0)l_0(x) + f(x_1)l_1(x) + \cdots + f(x_n)l_n(x)$$

where $l_0, \ldots, l_n$ are the Lagrange cardinal functions for the points $x_0, \ldots, x_n$, introduced in Chapter 2 at (2.2.2). Integrating over $[-1, 1]$ we get

$$\int_{-1}^{1} p_n(x)\, dx = \sum_{i=0}^{n} f(x_i) \int_{-1}^{1} l_i(x)\, dx = \sum_{i=0}^{n} f(x_i) w_i,$$

with

$$w_i = \int_{-1}^{1} l_i(x)\, dx,$$

These constants w_i are called the weights.

For example, with $n = 1$, we get the trapezoid rule. With $n = 2$, we get Simpson's rule. With $n = 3$, we get the Simpson's 3/8 rule

$$\int_{-1}^{1} f(x)\, dx \approx \frac{1}{4}\left[f(x_0) + 3f(x_1) + 3f(x_2) + f(x_3)\right].$$

With $n = 4$, we get the Boole's rule (or Milne's rule)

$$\int_{-1}^{1} f(x)\, dx \approx \frac{1}{45}\left[7f(x_0) + 32f(x_1) + 12f(x_2) + 32f(x_3) + 7f(x_4)\right].$$

It is easy to check that, for each given $n \geq 1$, the corresponding rule is exact for all polynomials of degree n.

Open Newton-Cotes formulas. If one uses a uniformly spaced grid, but not containing the endpoints $-1, 1$, this leads to the *open Newton-Cotes formulas*. Here we have

$$h = \frac{2}{n+2}, \qquad \tilde{x}_i = -1 + (i+1)h, \quad i = 0, 1, \cdots, n. \qquad (4.9.26)$$

The integration rule now has the form

$$\int_{-1}^{1} f(x)\, dx \approx \sum_{i=0}^{n} f(\tilde{x}_i)\tilde{w}_i,$$

where

$$\tilde{w}_i = \int_{-1}^{1} \tilde{l}_i(x)\, dx.$$

Here $\tilde{l}_0, \ldots, \tilde{l}_n$ are the cardinal function for the $n+1$ interpolating points $\tilde{x}_0, \ldots, \tilde{x}_n$ in (4.9.26). The degree of precision is at least n. Open formulas are less commonly used than the closed ones.

In general, for any $n+1$ arbitrary points $x_0, x_1, \cdots, x_n$ (possibly not uniformly spaced), one can design an interpolation rule as

$$\int_{-1}^{1} f(x)\, dx \approx w_0 f(x_0) + w_1 f(x_1) + \cdots + w_n f(x_n),$$

with

$$w_i = \int_{-1}^{1} l_i(x)\, dx.$$

Here $\{l_0, \ldots, l_n\}$ are the cardinal functions for the points $\{x_0, \ldots, x_n\}$. It is clear that any interpolation rule based on $n+1$ points has degree of accuracy at least n.

Gaussian Quadrature. If we do not specify in advance the location of the points x_i but try to position them in the best possible way, this leads to Gaussian quadrature. We now discuss a somewhat more general situation. Consider a rule for the integral

$$\int_{-1}^{1} w(x)f(x)\,dx \approx w_0 f(x_0) + w_1 f(x_1) + \cdots + w_n f(x_n). \qquad (A)$$

Here $w(x) > 0$ is a weight function. ($w(x)$ could be 0 at isolated points.) So far, in all our discussion we have used $w(x) \equiv 1$. Then, setting $f(x) = l_i(x)$ easily leads to the formula for the weights w_i, i.e.,

$$w_i = \int_{-1}^{1} l_i(x)w(x)\,dx. \qquad (B)$$

And the rule is exact for polynomials of degree n.

Our goal is to design a rule such that (A) is exact for polynomials of degree $2n + 1$, i.e., the rule is exact for $f(x) = x^k$, for $k = 0, 1, 2, \cdots, 2n + 1$, by suitably choosing the points x_i.

We define an inner-product of two functions f and g,

$$(f, g) = \int_{-1}^{1} f(x)g(x)w(x)\,dx. \qquad (4.9.27)$$

This qualifies as an inner product provided that $w(x) > 0$.

Starting from the set $\{1, x, x^2, \ldots, x^{2n+1}\}$, one can carry out the Gram-Schmidt orthogonalization procedure and generate a family of orthogonal polynomials, say $\{p_0, p_1, p_2, \ldots, p_{2n+1}\}$. Here "orthogonal" is meant in connection with the inner product (4.9.27). More precisely, to construct these polynomials $p_0, p_1, p_2, \cdots$ we start by defining

$$p_0(x) = 1.$$

The polynomial p_1 is then defined as

$$p_1(x) = x - \frac{(p_0, x)}{(p_0, p_0)}p_0(x) = x - \frac{1}{2}\int_{-1}^{1} p_0(x)xw(x)\,dx,$$

while p_2 is defined as

$$p_2(x) = x^2 - \frac{(p_0, x^2)}{(p_0, p_0)}p_0(x) - \frac{(p_1, x^2)}{(p_1, p_1)}p_1(x)$$

$$= x^2 - \frac{1}{2}\int_{-1}^{1} p_0(x)x^2 w(x)\,dx - \frac{p_1(x)}{(p_1, p_1)}\int_{-1}^{1} p_1(x)x^2 w(x)\,dx.$$

By induction, performing the Gram-Schmidt orthogonalization procedure one obtains the remaining polynomials $p_3, p_4, \ldots$. We observe that, by construction, all these polynomials are monic, i.e., their leading coefficient is always 1. Moreover they are orthogonal to each other:

$$(p_i, p_j) = \int_{-1}^{1} p_i(x)p_j(x)w(x)\,dx = 0, \quad \text{for all } i \neq j.$$

The set of polynomials $\{p_0, p_1, \cdots, p_n\}$ forms a basis for $\mathcal{P}^n$. In particular, p_{n+1} is orthogonal to all polynomials of degree n, i.e.,

$$(p_{n+1}, g) = 0, \quad \forall g \in \mathcal{P}^n.$$

The next theorem relates the zeros of these polynomials with the Gaussian Quadrature formulas.

Theorem 4.1. *Let $\{p_i\}_{i=1}^{\infty}$ be a family of orthogonal polynomials w.r.t. the weight function $w(x)$ over $[-1, 1]$. Let $x_0, x_1 \cdots, x_n$ be the zeros of p_{n+1}. Then the quadrature rule (A) with w_i defined in (B) is exact for $f(x) = x^k$, $k = 0, 1, \cdots, 2n+1$.*

Proof. Since $x_0, x_1 \cdots, x_n$ are the zeros of p_{n+1}, we have

$$p_{n+1}(x_i) = 0, \quad i = 0, 1, 2, \cdots, n.$$

By construction, we already know that the rule is exact for x^k with $k \leq n$. Now suppose that $f(x)$ is a polynomial of degree at most $2n + 1$. Then

$$f(x) = Q(x)p_{n+1}(x) + R(x),$$

where both $Q(x)$ and $R(x)$ are polynomials of degree at most n. One can view $Q(x)$ as the quotient upon synthetic division by $p_{n+1}(x)$ and $R(x)$ is the remainder. The rule applied to f yields

$$\sum_{i=0}^{n} w_i f(x_i) = \sum_{i=0}^{n} w_i \left(Q(x_i)p_{n+1}(x_i) + R(x_i) \right) = \sum_{i=0}^{n} w_i R(x_i),$$

because $p_{n+1}(x_i) = 0$. Then

$$\sum_{i=0}^{n} w_i R(x_i) = \int_{-1}^{1} w(x)R(x)\,dx = \int_{-1}^{1} w(x)R(x)\,dx + \int_{-1}^{1} w(x)Q(x)p_{n+1}(x)\,dx,$$

since $(p_{n+1}, Q) = 0$ because $Q \in \mathcal{P}^n$. Combining these facts, we obtain

$$\sum_{i=0}^{n} w_i f(x_i) = \int_{-1}^{1} w(x)R(x)\,dx + \int_{-1}^{1} w(x)Q(x)p_{n+1}(x)\,dx$$

$$= \int_{-1}^{1} w(x)[R(x) + Q(x)p_{n+1}(x)]\,dx = \int_{-1}^{1} w(x)f(x)\,dx,$$

showing that the rule is exact for any polynomial of degree $2n + 1$. $\qquad\square$

These orthogonal polynomials with various weight functions $w(x)$ have been extensively studied in the literature. Many of them are solutions to eigenvalue problems for two-point boundary value problems. Famous examples includes the Legendre equation (Legendre polynomials), Chebyshev equations (Chebyshev polynomials), Laguerre equations (Laguerre polynomials) etc. A detailed discussion of these issues is outside the scope of this book.

4.10   Homework Problems for Chapter 4

1. Trapezoid and Simpson's Methods

Given function $f(x) = e^{-x}$, we study different numerical approximations to the integral

$$\int_{0.0}^{0.8} f(x)\,dx.$$

We will use the values of $f(x)$ at the points $0.0, 0.2, 0.4, 0.6, 0.8$. Generate the data set before you start the numerical integration. Use 6-digits accuracy.

(a). Write out the trapezoid rule and compute the numerical integration with 6 digits.

(b). Write out the Simpson's rule and compute the numerical integration with 6 digits.

(c). What is the exact value of the integral? What is the absolute error by using trapezoid and Simpson's rule? Which method is better?

(d). The error formula for the trapezoid rule with $n + 1$ points yields

$$E_T(f; h) = -\frac{b - a}{12} h^2 f''(\xi), \quad h = \frac{b - a}{n},$$

for some $\xi \in (a, b)$. The error for Simpson's rule with $(2n + 1)$ points yields

$$E_S(f; h) = -\frac{b - a}{180} h^4 f^{(4)}(\xi'), \quad h = \frac{b - a}{2n},$$

for some $\xi' \in (a, b)$. If we wish the absolute value of the error to be smaller than 10^{-4}, how many points would be needed for each method?

2. Trapezoid Rule in Matlab

Preparation: Use the `help` in Matlab to learn how to use the function `feval`.

Write a Matlab function which computes the integral by trapezoid rule. Your function should be used by the following command in Matlab command window:

```
>> v=trapezoid('funItg',a,b,n)
```

where `funItg.m` is the name of the file of the function $f(x)$, and `a,b` is the interval, and `n` is the number of sub-intervals (i.e., $n + 1$ will be the number of points).

Test your function on problem 1. Write a script that computes for $n = 4, 8, 16, 32, 64, 128$. Compute also the absolute error for each n, and make a plot of the absolute error against n. (Use `loglog` to plot.) How does the error change when n is doubled? Do you expect this from the error estimate? Write your comments.

What to hand in?   Print out your function in the file `trapezoid.m`, your script file, the plot of error, and your comments.

3. Simpson's Rule in Matlab

Write a Matlab function which computes the integral by Simpson's rule. Your function should be used by the following command in Matlab command window:

```
>> v=Simpson('funItg',a,b,n)
```

where funItg.m is the name of the file of the function $f(x)$, and a, b is the interval, and n is the number of sub-intervals (i.e., $2n+1$ will be the number of points).

Test your function on problem 1. Write a script that computes, for $n = 2, 4, 8, 16, 32, 128$, the absolute error for each n, and a plot (with loglog) of the absolute error against n. How does the error change when n doubled? Compare the results with trapezoid rule and comment.

What to hand in? Your function in Simpson.m, your script file, the plot of error, and your comments.

4. Romberg Algorithm in Matlab

Preparation: Use helpdesk in Matlab to learn how to use the functions feval and quad.

(a). Write a Matlab function that computes the Romberg integration. One should be able to call the function by:

```
>> R=romberg('f',a,b,n)
```

where f is the name of the function where $f(x)$ is implemented, and a and b defines the integrating interval, and n is the size of your Romberg table. The function should return the whole Romberg table. The best approximation of the interval would be the value in R(n,n).
You may follow the pseudo-code in the lecture notes, or implement your own.
You may check your code against the result of the simulation in Section 4.8, to make sure that your code works.

(b). Use your **romberg** to compute the integrals

i) $$\int_0^\pi \sin(x)\,dx \qquad (= 2)$$

ii) $$\int_0^1 \sqrt{x}\,dx \qquad (= 2/3)$$

Compute also the errors. The exact values of the integrals are given in the parentheses above. Print the errors along the diagonal of the table, and note how it changes along the diagonal of the table. Use **format short e** in Matlab to display the error data.

(c). Explain why Romberg algorithm works poorly for the last integral.

(d). Use Matlab functions `quad` and `quadl` to compute the integrals in b). Use `1e-9` as tolerance for both integrations. Mark your observation.

What to hand in? The Matlab file `romberg.m`, the file for your function `f.m`, a script file that does b) and d), and the Romberg tables your get in b).

5. Numerical Integration and Extrapolation

Consider the function

$$f(t) = \begin{cases} \dfrac{\sin t}{t}, & t \neq 0, \\ 1, & t = 0. \end{cases}$$

Note that f is a continuous function for all t. Let

$$I(x) = \int_0^x f(t)\, dt \quad \text{and} \quad J = I(1).$$

(a). Compute approximations to J by trapezoid rule with 1,3 and 9 equal intervals. Try to use fewest possible time of computing $f(t)$. Show which formula you use.

(b). Derive a Richardson extrapolation algorithm which uses trapezoid approximations from a), to obtain better approximations to J. Show which formula you use. Generate the corresponding Romberg triangle.

6. Gaussian Quadrature and Beyond

(a). Consider the Gaussian Quadrature rule with 4 points on the interval $[-1,1]$,

$$\int_{-1}^{1} f(x)\, dx \approx a_1 f(x_1) + a_2 f(x_2) + a_3 f(x_3) + a_4 f(x_4)$$

where

$$x_1 = -\sqrt{\frac{1}{7}(3 - 4\sqrt{0.3})},$$

$$x_2 = -\sqrt{\frac{1}{7}(3 + 4\sqrt{0.3})},$$

$$x_3 = \sqrt{\frac{1}{7}(3 - 4\sqrt{0.3})},$$

$$x_4 = \sqrt{\frac{1}{7}(3 + 4\sqrt{0.3})},$$

where

$$a_1 = \frac{1}{2} + \frac{1}{12}\sqrt{\frac{10}{3}},$$

$$a_2 = \frac{1}{2} - \frac{1}{12}\sqrt{\frac{10}{3}},$$

$$a_3 = \frac{1}{2} + \frac{1}{12}\sqrt{\frac{10}{3}},$$

$$a_4 = \frac{1}{2} - \frac{1}{12}\sqrt{\frac{10}{3}}.$$

Show that the rule is exact for all polynomials of degree ≤ 7.

(b). Construct a rule of the form

$$\int_{-1}^{1} f(x)\, dx \approx a_1 f(-0.5) + a_2 f(0) + a_3 f(0.5)$$

that is exact for all polynomials of degree ≤ 2; that is, determine the values for a_1, a_2, a_3.

Chapter 5

Numerical Solutions of Non-linear Equations

5.1 Introduction

In this chapter, we study numerical methods for finding the zeros of a function $f(x)$, typically a non-linear function.

Problem description: $f(x)$ is a given function, real-valued, possibly non-linear. We want to find a root of f, that is: a point r such that $f(r) = 0$.

For any given non-linear function $f(x)$, there might be multiple roots, a unique root, or no roots at all. In general, roots of non-linear functions are very hard (or impossible) to find analytically. Below we give some simple examples to illustrate these ideas.

Example 5.1. Consider a quadratic polynomial $f(x) = x^2 + 5x + 6$. One can factorize it as

$$f(x) = (x + 2)(x + 3).$$

We thus have two roots: $r_1 = -2$ and $r_2 = -3$. Therefore, in this case roots are not unique.

Example 5.2. Consider now the polynomial $f(x) = x^2 + 4x + 10$. We have

$$f(x) = x^2 + 4x + 10 = (x + 2)^2 + 6 > 0.$$

There is no real number r such that $f(r) = 0$. Therefore, in this case roots do not exist.

Example 5.3. Consider a general non-linear function, for example

$$f(x) = x^2 - 3\cos x + e^x + \sqrt{x^2 + 1}.$$

In this case, one could prove that some roots exist, but it is impossible to find them analytically.

Our task: Find an approximation to a root, by some numerical method.

In this chapter we shall learn the following methods, and discuss their strengths and weaknesses:

- Bisection method;
- Fixed point iteration:
- Newton's method;
- Secant method.

5.2   Bisection Method

Consider a continuous function $f(x)$, for which we want to find a root r.

Basic idea: If we find two points a and b such that $f(a)$ and $f(b)$ have opposite sign, then by the *Intermediate Value Theorem* there exists a point c between a and b, such that $f(c) = 0$. This fact follows from the continuity of f. One can guess that the mid-point of the interval $[a, b]$ is a good candidate for the approximate solution.

We can iterate this procedure several times, in order to obtain a sufficiently accurate approximation. This leads to the *bisection method*.

Procedure: Here is the basic procedure of the bisection method:

---

(1) Initialization: evaluate the function f at random sampling points, until you find two values a, b such that $f(a) \cdot f(b) < 0$.

(2) Let $c = \frac{a+b}{2}$ be the mid-point of the interval $[a, b]$. If $f(c) = 0$, we are done. Otherwise:
if $f(c) \cdot f(a) < 0$ then pick the interval $[a, c]$,
if $f(c) \cdot f(b) < 0$, then pick the interval $[c, b]$.

(3) Apply again the procedure in step (2), with $[a, b]$ replaced by either $[a, c]$ or $[c, b]$ in the above two cases, until a stopping criteria is satisfied.

The students are encouraged to draw a picture and visualize this procedure.

Stop criteria. Various stopping criteria are suitable. Let $\epsilon > 0$ be the error tolerance. This is a small number that we choose in advance. We may decide to terminate the algorithm when any of the following happens:

1) The new subinterval has length $\leq \epsilon$.
2) The value $|f(c)|$ is very small, i.e., $|f(c)| \leq \epsilon$.
3) A maximum number of iterations is reached.
4) Any combination of the previous ones would work.

Convergence analysis. Consider $[a_0, b_0]$, $c_0 = \frac{a_0 + b_0}{2}$, let $r \in (a_0, b_0)$ be a root. The error e_0 has the estimate

$$e_0 = |r - c_0| \le \frac{b_0 - a_0}{2}.$$

We denote the further intervals as $[a_n, b_n]$ for iteration number n. Then we have

$$b_n - a_n = \frac{1}{2}(b_{n-1} - a_{n-1}), \qquad b_n - a_n = \frac{1}{2^n}(b_0 - a_0),$$

thus the error e_n can now be estimated as

$$e_n = |r - c_n| \le \frac{b_n - a_n}{2} \le \frac{b_0 - a_0}{2^{n+1}} = \frac{e_0}{2^n}.$$

If the error tolerance is ε, we can find out how many steps must be performed to achieve the desired accuracy. Since we require $e_n \le \varepsilon$, this leads to a condition on the number of steps n, namely

$$\frac{b_0 - a_0}{2^{n+1}} \le \varepsilon \qquad \Longrightarrow \qquad n \ge \frac{\ln(b-a) - \ln(2\varepsilon)}{\ln 2}. \tag{5.2.1}$$

Remark: (i) We observe that this is a very slow convergence. (ii) The estimate (5.2.1) depends only on the interval (a, b), and it does not depend on the function f. We only need to know that f has opposite signs at a and b.

5.3   Fixed Point Iterations

Main idea. We rewrite the equation $f(x) = 0$ in the form $x = g(x)$, for a suitable function g. Then, the root r of the function f now becomes the *fixed point* for the function g. That means:

$$f(r) = 0 \qquad \text{if and only if} \qquad r = g(r). \tag{5.3.2}$$

One can then use iterations to find this fixed point.

Remark. There are many ways to choose the function g so that (5.3.2) holds. For example, one could write

$$g(x) = f(x) + x.$$

Another way could be

$$g(x) = x - f(x).$$

A smart choice of g can make a big difference in the convergence of the algorithm.

Fixed point iteration algorithm:

- Choose a starting point x_0,
- Compute $x_{k+1} = g(x_k)$, $k = 0, 1, 2, \ldots$, until a stopping criterion is met.

Stop Criteria: Let ε be the tolerance. This is a small number, chosen in advance. Any combination of the following stopping criteria can be used:

- $|x_k - x_{k-1}| \leq \varepsilon$;
- $|x_k - g(x_k)| \leq \varepsilon$;
- A maximum number of iterations is reached.

We first show a successful example.

Example 5.4. Find an approximate solution to

$$f(x) = x - \cos x = 0,$$

using 4-digit accuracy.

Answer. Since we have $x = \cos x$, we can choose $g(x) = \cos x$. The iteration step is

$$x_{k+1} = \cos(x_k).$$

Choosing the initial guess $x_0 = 1$, we can perform the above iteration many times, using a calculator. Here is what we get:

$$x_1 = \cos x_0 = 0.5403$$
$$x_2 = \cos x_1 = 0.8576$$
$$x_3 = \cos x_2 = 0.6543$$
$$\vdots$$
$$x_{23} = \cos x_{22} = 0.7390$$
$$x_{24} = \cos x_{23} = 0.7391$$
$$x_{25} = \cos x_{24} = 0.7391$$

We see that we can stop here, since all 4 digits are unchanged after the last iteration. Our approximation to the root is 0.7391.

We now look at a not-so-fortunate example.

Example 5.5. Consider the function

$$f(x) = e^{-2x}(x - 1) = 0.$$

We see that $x = r = 1$ is a root.

We can rewrite the equation $f(x) = 0$ as a fixed point problem

$$x = g(x) = e^{-2x}(x - 1) + x.$$

This gives us the iteration scheme

$$x_{k+1} = g(x_k) = e^{-2x_k}(x_k - 1) + x_k.$$

We choose an initial guess $x_0 = 0.99$, which is very close to the real root $r = 1$. Do you think we shall get closer to the root as we iterate? Here are the results:

$$x_1 = g(x_0) = 0.9886$$
$$x_2 = g(x_1) = 0.9870$$
$$x_3 = g(x_2) = 0.9852$$

$$\vdots$$

$$x_{27} = g(x_{26}) = 0.1655$$
$$x_{28} = g(x_{27}) = -0.4338$$
$$x_{29} = g(x_{28}) = -3.8477.$$

It is now clear that the iteration diverges, without finding the root.

What went wrong?

Observation: The convergence of the fixed point iteration depends on the function $g(x)$. We illustrate this by two typical graphs, one with $|g'(x)| > 1$ and the other with $|g'(x)| < 1$ on an interval around the fixed point r. See Figure 5.1 for an illustration. The intersection point of the graphs of $y = x$ and $y = f(x)$ gives the fixed point for $g(x)$.

When $|g'(x)| < 1$ as in the left plot, the iteration converges, while the case $|g'(x)| > 1$ as in the right plot the iteration diverges. This suggests that the convergence depends on the size of the derivative $|g'(x)|$, for x near the root r.

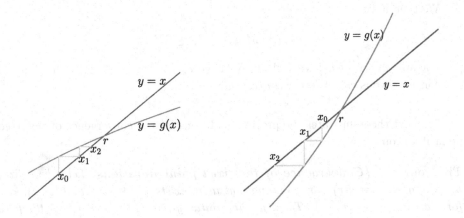

Fig. 5.1   Illustrations of fixed point iterations. Left: convergent case. Right: divergent case.

Convergence analysis. Let r be the fixed point for $g(x)$, so that $r = g(r)$.

Our iteration is $x_{k+1} = g(x_k)$.

We define the error at iteration step k as

$$e_k = |x_k - r|.$$

This error evolves at each iteration. It can be estimated by

$$
\begin{aligned}
e_{k+1} &= |x_{k+1} - r| \\
&= |g(x_k) - r| \\
&= |g(x_k) - g(r)| \\
&= |g'(\xi) \cdot (x_k - r)| \qquad \text{for some } \xi \in (x_k, r) \\
&= |g'(\xi)|\, e_k.
\end{aligned}
$$

We see that, after each iteration, the error is multiplied by the factor $|g'(\xi)|$, i.e.,

$$e_{k+1} = |g'(\xi)|\, e_k.$$

This leads to the following observation:

- If $|g'(\xi)| > 1$, then $e_{k+1} > e_k$. Hence the error increases at each step, and the iteration diverges.
- If $|g'(\xi)| < 1$, then $e_{k+1} < e_k$. Hence the error decreases at each step and the iteration converges. In this case, consider the upper bound

$$M = \sup_{\xi} |g'(\xi)| < 1.$$

We then have

$$e_{k+1} \leq M e_k, \qquad e_k \leq M^k e_0.$$

For any choice of e_0, we will always have $e_k \to 0$ as $k \to \infty$. This kind of convergence is called *linear convergence*.

The next theorem states the precise conditions for the convergence of the fixed point iteration.

Theorem 5.1. (Convergence of the fixed point iterations) *Let r be a fixed point of g, so that $g(r) = r$. Assume that there exists $a > 0$ such that $|g'(x)| < 1$ for every $x \in [r - a, r + a]$. Then, for any initial guess $x_0 \in [r - a, r + a]$ the fixed point iterations*

$$x_1 = g(x_0), \qquad x_2 = g(x_1), \qquad \cdots \qquad x_{n+1} = g(x_n), \qquad \cdots$$

converge to r as $n \to \infty$.

In particular, if $|g'(r)| < 1$, then these iterations converge for every initial guess x_0 sufficiently close to r.

In Example 5.4, we have

$$g(x) = \cos x, \qquad g'(x) = \sin x, \qquad r = 0.7391,$$

so

$$|g'(r)| = |\sin(0.7391)| < 1.$$

By continuity, for x in a small interval $[r - a, r + a]$ we still have $|g'(x)| < 1$. Therefore the iteration converges.

In Example 5.5, we have

$$g(x) = e^{-2x}(x - 1) + x,$$
$$g'(x) = -2e^{-2x}(x - 1) + x^{-2x} + 1.$$

At the fixed point $r = 1$, we have

$$|g'(r)| = e^{-2} + 1 > 1.$$

Here the convergence condition fails, and the iterations diverge.

Pseudo code:

```
r=fixedpoint('g', x,tol,nmax)
  r=g(x); % first iteration
  nit=1;
  while (abs(r-g(r))>tol and nit < nmax) do
    r=g(r);
    nit=nit+1;
  end
end
```

We will now try to answer the following question: Assume that the condition

$$|g'(x)| \leq M < 1, \qquad \text{for every} \quad x \in [r - a, r + a]$$

is satisfied. If we require certain error tolerance, for example, $e_n \leq \varepsilon$, what is the minimum number of iterations needed?

Answer. From the error analysis, we have

$$e_{k+1} \leq M e_k.$$

This gives

$$e_1 \leq M e_0,$$
$$e_2 \leq M e_1 \leq M^2 e_0,$$
$$\cdots$$
$$e_k \leq M^k e_0.$$

To use the above formula, we still need to estimate e_0. Note that we have

$$
\begin{aligned}
e_0 &= |r - x_0| \\
&= |r - x_1 + x_1 - x_0| \\
&\le e_1 + |x_1 - x_0| \\
&\le M e_0 + |x_1 - x_0|.
\end{aligned}
$$

Hence

$$
e_0 \le \frac{1}{1 - M} |x_1 - x_0|,
$$

where the right-hand side of the above expression can be computed! Putting everything together, we obtain

$$
e_k \le \frac{M^k}{1 - M} |x_1 - x_0|.
$$

If the error tolerance is ε, then we have the constraint

$$
e_k \le \frac{M^k}{1 - M} |x_1 - x_0| \le \varepsilon
$$

$$
\Rightarrow \quad M^k \le \frac{\varepsilon(1 - M)}{|x_1 - x_0|}.
$$

The number of iterations needed to achieve an error $e_k \le \varepsilon$ is thus

$$
k \ge \frac{\ln(\varepsilon(1 - M)) - \ln|x_1 - x_0|}{\ln M}. \tag{5.3.3}
$$

Example 5.6. We want to solve $\cos x - x = 0$ with the fixed point iteration

$$
x = g(x) = \cos x,
$$

using $x_0 = 1$. We know $r \approx 0.74$. We see that the points x_k computed by the iteration remain inside the interval $[0, 1]$. For $x \in [0, 1]$, we have

$$
|g'(x)| = |\sin x| \le \sin 1 = 0.8415 \doteq M
$$

And we also have

$$
x_1 = \cos x_0 = \cos 1 = 0.5403.
$$

Then, to achieve an error $\le \varepsilon = 10^{-5}$, using the formula (5.3.3), the minimum number of iterations needed is

$$
k \ge \frac{\ln(\varepsilon(1 - M)) - \ln|x_1 - x_0|}{\ln M} \approx 73.
$$

You should be aware that the above estimate gives an upper bound on the number of iterations, valid even in the worst case scenario. In practice, the actual simulations will often give you much better results. Give it a try in Matlab and you will find that $k = 28$ is enough.

Matlab Simulations. One can implement this simulation in Matlab by the following simple code.

```
x0=1;
x1=cos(x0);
tol=1e-5;
err=abs(x0-x1);
nmax=50;
nit=0;
while (nit<nmax & err >tol)
  x0=x1;
  x1=cos(x0);
  err=abs(x0-x1);
  nit=nit+1;
  disp(sprintf("n=%d, x=%f, error=%e",nit,x1,err));
end
```

And you will get the following output.

```
n=1, x=0.857553, error=3.172509e-01
n=2, x=0.654290, error=2.032634e-01
n=3, x=0.793480, error=1.391906e-01
n=4, x=0.701369, error=9.211159e-02
n=5, x=0.763960, error=6.259091e-02
n=6, x=0.722102, error=4.185726e-02
n=7, x=0.750418, error=2.831534e-02
n=8, x=0.731404, error=1.901372e-02
n=9, x=0.744237, error=1.283331e-02
n=10, x=0.735605, error=8.632614e-03
n=11, x=0.741425, error=5.820346e-03
n=12, x=0.737507, error=3.918196e-03
n=13, x=0.740147, error=2.640445e-03
n=14, x=0.738369, error=1.778131e-03
n=15, x=0.739567, error=1.197998e-03
n=16, x=0.738760, error=8.068823e-04
n=17, x=0.739304, error=5.435725e-04
n=18, x=0.738938, error=3.661357e-04
n=19, x=0.739184, error=2.466431e-04
n=20, x=0.739018, error=1.661373e-04
n=21, x=0.739130, error=1.119141e-04
n=22, x=0.739055, error=7.538578e-05
n=23, x=0.739106, error=5.078118e-05
n=24, x=0.739071, error=3.420663e-05
```

```
n=25, x=0.739094, error=2.304208e-05
n=26, x=0.739079, error=1.552138e-05
n=27, x=0.739089, error=1.045541e-05
n=28, x=0.739082, error=7.042881e-06
```

The algorithm stops here since the error is smaller than the tolerance.

5.4 Newton's Method

Assumption: Consider a function $f(x)$ in $\mathcal{C}^2$ (i.e., twice continuously differentiable), which has a root r such that
$$f(r) = 0, \qquad f'(r) \neq 0.$$
We will describe a more efficient method to approximate r, using iterations.

Graphic interpretation of Newton's method. Assume that, after k steps, you have computed an approximation x_k. To compute the next approximation x_{k+1}, we approximate $f(x)$ with a linear function near x_k. Then we let x_{k+1} be the zero of this linear approximation. See Figure 5.2 for an illustration.

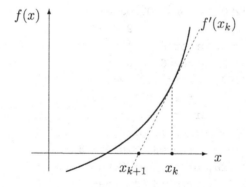

Fig. 5.2   Newton's method: linearize $f(x)$ at x_k and let x_{k+1} be the zero of this linear approximation.

The tangent line to the graph of f at the point x_k has equation
$$y = f(x_k) + f'(x_k)(x - x_k).$$
Calling x_{k+1} the point where this line crosses the x-axis, (i.e., where $y = 0$), we find
$$0 = f(x_k) + f'(x_k)(x_{k+1} - x_k).$$
Solving for x_{k+1} we obtain famous **Newton's iteration formula:**

$$x_{k+1} = x_k - \frac{f(x_k)}{f'(x_k)} = x_k - \Delta x_k, \qquad \Delta x_k = \frac{f(x_k)}{f'(x_k)}.$$

Newton's method viewed as a fixed point iteration. Newton's method can be written as a fixed point iteration, namely

$$x_{k+1} = g(x_k), \qquad \text{where} \qquad g(x) = x - \frac{f(x)}{f'(x)}.$$

If r is a fixed point, assuming that $f'(r) \neq 0$ we have

$$r = g(r) = r - \frac{f(r)}{f'(r)}, \qquad \frac{f(r)}{f'(r)} = 0, \qquad f(r) = 0.$$

Hence r is a root of f.

The derivative of g is computed by

$$g'(x) = 1 - \frac{f'(x)f'(x) - f''(x)f(x)}{(f'(x))^2} = \frac{f''(x)f(x)}{(f'(x))^2}.$$

In particular, at $x = r$ we have

$$g'(r) = \frac{f''(r)f(r)}{(f'(r))^2} = 0.$$

This is the best possible scenario for a fixed point iteration! Indeed, by our previous error analysis, at each iteration the error is reduced by a factor of $|g'(x)|$, for x in a small neighborhood of r. Having $g'(r) = 0$ is the best we can hope for, in order to achieve fast convergence.

We claim that, in a way, Newton's iteration is the "best possible" fixed point iteration. To see this, observe that the equation

$$f(x) = 0,$$

can be written in the equivalent form

$$b(x)f(x) = 0,$$

(with $b(x) \neq 0$), and then as a fixed point equation

$$x = x - b(x)f(x) = g(x).$$

The idea is now to choose the function $b(x)$ in a smart way, achieving the fastest possible convergence. This will be the case if $g'(r) = 0$.

Computing this derivative, we find

$$g'(x) = 1 - b'(x)f(x) - b(x)f'(x).$$

At the point r, we have $f(r) = 0$ and hence

$$g'(r) = 1 - b(r)f'(r).$$

Choosing $b(x) = \frac{1}{f'(x)}$ we obtain $g'(r) = 0$, as we wanted. With this choice, the fixed point iteration takes the form

$$x = g(x) = x - b(x)f(x) = x - \frac{f(x)}{f'(x)},$$

which is precisely Newton's iteration formula.

Convergence analysis. We have the following Theorem.

Theorem 5.2. (*Quadratic error estimates for Newton's iterations*) *Let $f(x)$ be a smooth function and let r be a root, so that $f(r) = 0$. Assume that $f'(r) \neq 0$ and consider the Newton iterations*

$$x_{k+1} = x_k - \frac{f(x_k)}{f'(x_k)}.$$

Then there is a constant K such that for all x_k, x_{k+1} sufficiently close to r, the error $e_k = |x_k - r|$ satisfies

$$e_{k+1} \leq K(e_k)^2$$

The above rate of convergence is called **quadratic convergence.**

Proof. Let

$$g(x) = x - \frac{f(x)}{f'(x)}.$$

Let r be the root so $f(r) = 0$ and $r = g(r)$. Define the error:

$$e_{k+1} \doteq |x_{k+1} - r| = |g(x_k) - g(r)|.$$

The Taylor approximation for $g(x_k)$ at r gives

$$g(x_k) = g(r) + (x_k - r)g'(r) + \frac{1}{2}(x_k - r)^2 g''(\xi), \qquad \text{for some} \quad \xi \in (x_k, r).$$

The derivatives of g are

$$g'(x) = \frac{f''(x)f(x)}{(f'(x))^2},$$

$$g''(x) = \frac{f'(x)f''(x) + f(x)f'''(x)}{[f'(x)]^2} - \frac{2f(x)[f''(x)]^2}{[f'(x)]^3}.$$

Recalling that $f(r) = 0, f'(r) \neq 0$, at $x = r$ we find

$$g'(r) = 0,$$

$$g''(r) = \frac{f''(r)}{f'(r)}.$$

Inserting these values in the Taylor approximation of $g(x_k)$, we obtain

$$g(x_k) = g(r) + \frac{1}{2}(x_k - r)^2 g''(\xi).$$

The error can be estimated as

$$e_{k+1} = |x_{k+1} - r| = |g(x_k) - g(r)| = \frac{1}{2}(x_k - r)^2 |g''(\xi)| = \frac{|g''(\xi)|}{2} e_k^2.$$

Since g is a smooth function, its second derivative is a continuous function. Hence there exists a constant K such that

$$\frac{|g''(\xi)|}{2} \leq K$$

for all points ξ sufficiently close to r. The previous inequality yields

$$e_{k+1} \leq K(e_k)^2,$$

proving the quadratic convergence. $\qquad\qquad\qquad\qquad\qquad\qquad\qquad\qquad\square$

Remark: The assumptions that $f'(r) \neq 0$ and $f''(r)$ is bounded are crucial in the proof. If they are not satisfied, then the conclusion of the theorem may fail, and Newton's iteration will not achieve quadratic convergence. Take for example $f(x) = x^2$. In this case $r = 0$, while $f'(x) = 2x$ and $f'(r) = f'(0) = 0$. The Newton iteration takes the form

$$x_{k+1} = x_k - \frac{f(x_k)}{f'(x_k)} = x_k - \frac{x_k^2}{2x_k} = \frac{1}{2}x_k.$$

In this case the error is $e_k = |x_k - r| = |x_k|$, and it satisfies

$$e_{k+1} = \frac{1}{2}e_k.$$

In this case, $f'(r) = 0$ and we only achieve linear convergence.

Can you think of an example for the case where $f''(x)$ is unbounded for x close to the root r?

The next theorem shows that, if the error e_k satisfies the quadratic estimates in Theorem 5.2, then it actually converges to zero, provided that the initial error e_0 is sufficiently small. Notice that here the convergence is achieved even if the constant K is very large.

Theorem 5.3. *(Quadratic error estimates $\implies$ convergence) Assume that, at each iteration, the error satisfies*

$$e_{k+1} \leq K \left(e_k\right)^2 \tag{5.4.4}$$

for some constant K (possibly large). Then, if e_0 is sufficiently small, we have the convergence $e_k \to 0$ as $k \to \infty$

Proof. Let e_0 be small enough, so that $Ke_0 < 1$.
Using the quadratic bound (5.4.4) with $k = 1, 2, 3, \ldots$, we obtain

$$e_1 \leq (Ke_0)e_0 < e_0,$$

$$e_2 \leq (Ke_1)e_1 \leq (Ke_0)e_1 \leq (Ke_0)^2 e_0 < e_0,$$

$$e_3 \leq (Ke_2)e_2 \leq (Ke_0)(Ke_0)^2 e_0 = (Ke_0)^3 e_0.$$

Continuing by induction we see that, for every $k \geq 1$, one has

$$e_k \leq (Ke_0)^k e_0. \tag{5.4.5}$$

Since $Ke_0 < 1$, this implies

$$\lim_{k \to \infty} e_k \leq \lim_{k \to \infty} (Ke_0)^k e_0 = 0.$$

$\square$

Example 5.7. Find a numerical method to compute $\sqrt{a}$ using only $+, -, *, /$ arithmetic operations. Test it for $a = 3$.

Answer. We observe that $\sqrt{a}$ is a root of the polynomial $f(x) = x^2 - a$. Newton's method gives

$$x_{k+1} = x_k - \frac{f(x_k)}{f'(x_k)} = x_k - \frac{x_k^2 - a}{2x_k} = \frac{x_k}{2} + \frac{a}{2x_k}.$$

We test it on $a = 3$. As initial point, we choose $x_0 = 1.7$.

approximation	error
$x_0 = 1.7$	3.2×10^{-2},
$x_1 = 1.7324$	3.0×10^{-4},
$x_2 = 1.7321$	2.6×10^{-8},
$x_3 = 1.7321$	4.4×10^{-16}.

Note the extremely fast convergence. Are you surprised or did you expect it? Does it look reasonable? Indeed, if e_0 is of order 10^{-2}, then e_1 will be of order $(e_1)^2 = 10^{-4}$, and so on and so forth. Usually, if the initial guess is good (i.e., close to r), then only a couple of iterations are enough to get a very accurate approximation.

Stop criteria: Given an error tolerance $\varepsilon > 0$, the Newton algorithm can be iterated until any one of the following conditions is met:

- $|x_k - x_{k-1}| \leq \varepsilon$,
- $|f(x_k)| \leq \varepsilon$,
- a (predetermined) number of iterations is reached.

Sample Pseudo-Code:

```
r=newton('f','df',x0,nmax,tol)
x=x0; n=0; dx=f(x)/df(x);
while ((dx>tol) and (f(x)>tol)) or (n<nmax) do
   n=n+1;
   x=x-dx;
   dx=f(x)/df(x);
end
r=x-dx;
```

5.5   Secant Method

For some function $f(x)$ having a complicated form, the derivative $f'(x)$ might not be available, or very expensive to compute. For this reason, it is convenient to use an iteration algorithm that does not require the computation of the derivative $f'(x)$.

This can be achieved by replacing f' with a finite difference approximation, namely

$$f'(x_k) \approx \frac{f(x_k) - f(x_{k-1})}{x_k - x_{k-1}}.$$

This idea leads to the *secant method*:

$$x_{k+1} = x_k - \frac{x_k - x_{k-1}}{f(x_k) - f(x_{k-1})} f(x_k).$$

Note that here the algorithm computes the next point x_{k+1}, given the previous two points x_{k-1}, x_k. To start the iteration, one has to provide two initial guesses: x_0, x_1.

Advantages of secant method:

- No computation of the derivative f' is required;
- At each step, the function $f(x)$ must be computed only at one new point;
- The algorithm achieves fast convergence.

Convergence result: In the same situation considered in Theorem 5.2, one can show that, for the secant method, the errors satisfy

$$e_{k+1} \leq K(e_k)^\alpha, \qquad \text{where} \qquad \alpha = \frac{1}{2}(1 + \sqrt{5}) \approx 1.62,$$

for some constant K. This is called *superlinear convergence*, since $1 < \alpha < 2$, so the convergence is faster than linear convergence, but slower than quadratic convergence.

The secant method converges for any smooth function f, provided that the two initial points x_0 and x_1 are sufficiently close to the root r, and $f'(r) \neq 0$.

Example 5.8. Use the secant method for computing $\sqrt{a}$.

Answer. Let $f(x) = x^2 - a$. The secant iteration algorithm is

$$x_{k+1} = x_k - \frac{(x_k^2 - a)(x_k - x_{k-1})}{(x_k^2 - a) - (x_{k-1}^2 - a)}$$

$$= x_k - \frac{x_k^2 - a}{x_k + x_{k-1}}.$$

We now test the secant method for $a = 3$, with initial guesses $x_0 = 1.65$, $x_1 = 1.7$. We get the following results:

approximation	error
$x_1 = 1.7$	$e_1 = 3.2 \times 10^{-2}$,
$x_2 = 1.7328$	$e_2 = 7.9 \times 10^{-4}$,
$x_3 = 1.7320$	$e_3 = 7.3 \times 10^{-6}$,
$x_4 = 1.7321$	$e_4 = 1.7 \times 10^{-9}$,
$x_5 = 1.7321$	$e_5 = 3.6 \times 10^{-15}$.

It is a bit slower than Newton's method, but not much. The convergence is still very fast.

The most practical algorithm: hybrid methods. Such methods involve two steps:

- Use the bisection method, maybe with 5-6 iterations, to get some good initial guess x_0;
- Use either Newton or the secant method, with this x_0, for 3-4 iterations.

5.6   Systems of Non-linear Equations

The same techniques discussed in this chapter can be used to find zeros of vector valued functions, with several variables. For example, consider the system of equations

$$\begin{cases} f_1(x_1, x_2, \cdots, x_n) = 0, \\ f_2(x_1, x_2, \cdots, x_n) = 0, \\ \qquad\vdots \\ f_n(x_1, x_2, \cdots, x_n) = 0. \end{cases} \tag{5.6.6}$$

Adopting vector notation, this can be written as

$$\mathbf{F}(\vec{x}) = 0,$$

where

$$\mathbf{F} = (f_1, f_2, \cdots, f_n), \qquad \vec{x} = (x_1, x_2, \cdots, x_n).$$

Fixed point iteration. We first find a function $\mathbf{G}$ such that

$$\mathbf{F}(\vec{x}) = 0 \qquad \text{if and only if} \qquad \vec{x} = \mathbf{G}(\vec{x}).$$

To solve the equivalent equation $\vec{x} = \mathbf{G}(\vec{x})$, we choose an initial guess $\vec{x}_0$, and construct the sequence of vectors

$$\vec{x}_{k+1} = \mathbf{G}(\vec{x}_k).$$

Under suitable assumptions one can prove that, as $k \to \infty$, the vectors $\vec{x}_k \in \mathbb{R}^n$ converge to a solution of the system (5.6.6).

Newton's method: When $\mathbf{F}$ is a function of several variables, the Newton iteration formula becomes:

$$\vec{x}_{k+1} = \vec{x}_k - D\mathbf{F}(\vec{x}_k)^{-1} \cdot \mathbf{F}(\vec{x}_k).$$

Here $D\mathbf{F}(\vec{x}_k)$ is the Jacobian matrix of first-order partial derivatives of $\mathbf{F} = (f_1, \ldots, f_n)$, namely

$$D\mathbf{F}(\vec{x}_k) = \begin{pmatrix} \dfrac{\partial f_1}{\partial x_1} & \dfrac{\partial f_1}{\partial x_2} & \cdots & \dfrac{\partial f_1}{\partial x_n} \\[2mm] \dfrac{\partial f_2}{\partial x_1} & \dfrac{\partial f_2}{\partial x_2} & \cdots & \dfrac{\partial f_2}{\partial x_n} \\[2mm] \vdots & \vdots & \ddots & \vdots \\[2mm] \dfrac{\partial f_n}{\partial x_1} & \dfrac{\partial f_n}{\partial x_2} & \cdots & \dfrac{\partial f_n}{\partial x_n} \end{pmatrix},$$

where all derivatives are computed at the point $\vec{x}_k \in \mathbb{R}^n$. Moreover, $D\mathbf{F}(\vec{x}_k)^{-1}$ is the inverse matrix of $D\mathbf{F}(\vec{x}_k)$.

A detailed discussion of the error estimates and of the convergence of these methods is outside the scope of these lecture notes.

5.7   Homework Problems for Chapter 5

1. Bisection Method

We consider the bisection method to find a root for $f(x) = 0$ where f is a continuous function. If $f(0) < 0$ and $f(1) > 0$, then there is a root on the interval $[0, 1]$, and we take $[0, 1]$ as the initial interval. How many steps of the bisection method are needed to obtain an approximation to the root, if the absolute error is guaranteed to be $\leq 10^{-8}$? (Note that your answer does not depend on the actual function!)

2. Fixed Point Iteration

Given a function $f(x) = e^{-x} - \cos(x)$. We want to find a root r such that $f(r) = 0$.

(a). First show that there is a root inside the interval $[1.1, 1.6]$.

(b). Next, we try to locate this root by a fixed point iteration. We observe that $f(x) = 0$ is equivalent to

$$x = g_1(x), \qquad \text{where} \qquad g_1(x) = f(x) + x.$$

Set up the corresponding fixed point iteration scheme. Then, choose the starting point $x_0 = 1.6$, and perform 4 iterations to compute the values x_1, x_2, x_3, x_4. What do you observe? Does the method converge? Why?

(c). We now observe that $f(x) = 0$ is also equivalent to

$$x = g_2(x), \qquad \text{where} \qquad g_2(x) = x - f(x).$$

Based on this new function, set up the corresponding fixed point iteration. Choose again $x_0 = 1.6$ as your starting point, and perform 4 iterations to compute the values x_1, x_2, x_3, x_4. What is your result? How different is it from part (b)? Does the method converge? Why?

3. More on Fixed Point Iteration

Consider the iteration scheme:

$$x_{n+1} = \frac{1}{2}x_n + \frac{1}{x_n}, \qquad x_0 = 1.$$

(a). If the iteration converges, what is $\lim_{n\to\infty} x_n$?

(b). Does the method converge? Why?

4. Newton's Method

As you have seen in Example 5.7 in Section 5.4, the computation of $\sqrt{R}$ can be carried out by finding the root of $f(x) = x^2 - R$ using Newton's method.

(a). Check that in this case Newton's iteration scheme is

$$x_{n+1} = \frac{1}{2}\left(x_n + \frac{R}{x_n}\right).$$

(b). Show that the sequence $(x_n)_{n\geq 1}$ satisfies

$$x_{n+1}^2 - R = \left[\frac{x_n^2 - R}{2x_n}\right]^2.$$

Interpret this equation in terms of quadratic convergence.

The quantity $x_n^2 - R = f(x_n)$ is also called the *residual*. It gives a measure of the error in the approximation x_n.

(c). Now we test the iterations for $R = 10$. Choose $x_0 = 3$, and compute x_1, x_2, x_3 and x_4. Use 8 digits in your computation. Comment on your result.

5. When Newton's Method Does not Work Well

Let m be a positive integer and consider the function

$$f(x) = (x - 1)^m,$$

It is clear that $r = 1$ is a root (indeed, it is a root with multiplicity m).

We now try to find this root by Newton's iteration, for $m = 8$.

(a). Set up the iteration scheme.

(b). Use $x_0 = 1.1$ as your initial guess (note that this is very close to the root $r = 1$). Complete 4 iterations to compute the values x_1, x_2, x_3, x_4.

(c). How does the method work for this problem? Do we still have quadratic convergence? Can you explain what is causing this?

(d). If $m = 20$ (or with any large value of m), can you predict the behavior of Newton's iteration? What type of convergence will you get? Explain in detail.

6. Newton's Method in Matlab

Preparation: Use "help sprintf" and "help disp" in Matlab to understand how to use "sprintf" and "disp" to display the data. Here is an example:

```
disp(sprintf('I have n=%d and x=%g  but f=%f.\n',2,1.22, 1.22))
```

This will give the following result in Matlab:

```
I have n=2 and x=1.22  but f=1.220000.
```

The problem: Write a Matlab function for Newton's method. Your file mynewton.m should start with:

```
function x=mynewton(f,df,x0,tol,nmax)
```

Among the input variables, `f,df` are the function f and its derivative f', `x0` is the initial guess, `tol` is the error tolerance, and `nmax` is the maximum number of iterations. The output variable `x` is the result of the Newton iterations. Use `sprintf` and "disp" to display your result in each iteration so you can check the convergence along the way.

First, test your function with Example 5.7 in Section 5.4, computing $\sqrt{2}$.

Then, use your Matlab function to find a root of $f(x) = e^{-x} - \cos(x)$ on the interval $[1.1, 1.6]$. Use `tol=1e-12` and `nmax=10`. You should choose an initial guess x_0 on the interval $[1.1, 1.6]$. What is your answer?

What to hand in: Print out your files `mynewton.m`, files for functions $f(x)$ and $f'(x)$, script file, test result for the example of $\sqrt{2}$ and for the root of $f(x) = e^{-x} - \cos(x)$.

7. The Secant Method

a). Calculate an approximate value for $4^{3/5}$ using the secant method with $x_0 = 3$ and $x_1 = 2$. Make 3 steps, and compute the values x_2, x_3, x_4. Comment on your result.

b). Use secant method to find the root for $f(x) = x^3 - 2x + 1$ with $x_0 = 4$ and $x_1 = 2$. Compute the values x_2, x_3 and x_4. Does the method converge? (You can easily check that $x = 1$ is a root.)

c). Consider the iteration scheme

$$x_{n+1} = x_n + (2 - e^{x_n})(x_n - x_{n-1})/(e^{x_n} - e^{x_{n-1}}), \qquad x_0 = 0, \quad x_1 = 1.$$

If the iteration converges, what is $\lim_{n \to \infty} x_n$?

8. The Secant Method in Matlab

Write a Matlab function to locate a root by the secant method. Your function should be put in the file `mysecant.m`, and should be called as

```
x=mysecant(f,x0,x1,tol,nmax)
```

Here `f` is the function f, `x0,x1` are the initial guesses, `tol` is the error tolerance, `nmax` is the maximum number of iterations, and `x` is the output of your function.

Test your function to find the value $\sqrt{2}$, as the root for $f(x) = x^2 - 2$, to see if it works. Use `sprintf` and "disp" to display your result in each iteration, so you can check the convergence. Then, use it to find a root for the function $f(x) = e^{-x} - \cos(x)$ in the interval $[1.1, 1.6]$. Use `tol=1e-12` and `nmax=10`. Your two initial guesses should be on the interval $[1.1, 1.6]$. How does the result compare to those with Newton's method? Comment in detail.

What to hand in: Your file `mysecant.m`, test result for $f(x) = x^2 - 2$, and the result for $f(x) = e^{-x} - \cos(x)$.

Chapter 6

Direct Methods for Systems of Linear Equations

6.1 Introduction

In this chapter we study numerical methods to solve a system of linear equations

$$\begin{cases} a_{11}x_1 + a_{12}x_2 + \cdots + a_{1n}x_n = b_1, & (1) \\ a_{21}x_1 + a_{22}x_2 + \cdots + a_{2n}x_n = b_2, & (2) \\ \qquad\qquad\qquad\vdots & \\ a_{n1}x_1 + a_{n2}x_2 + \cdots + a_{nn}x_n = b_n. & (n) \end{cases} \qquad (6.1.1)$$

Here we have n equations and n unknowns.

Using the summation symbol, we can write (6.1.1) in the more compact form

$$\sum_{j=1}^{n} a_{ij}x_j = b_i, \qquad i = 1, \cdots, n.$$

One can also write (6.1.1) in matrix-vector form:

$$A\vec{x} = \vec{b},$$

where $A \in I\!\!R^{n\times n}$ is the *coefficient matrix*, $\vec{x} \in I\!\!R^n$ is the unknown vector, and $\vec{b} \in I\!\!R^n$ is the load vector:

$$A = \{a_{ij}\} = \begin{pmatrix} a_{11} & a_{12} & \cdots & a_{1n} \\ a_{21} & a_{22} & \cdots & a_{2n} \\ \vdots & \vdots & \ddots & \vdots \\ a_{n1} & a_{n2} & \cdots & a_{nn} \end{pmatrix}, \qquad \vec{x} = \begin{pmatrix} x_1 \\ x_2 \\ \vdots \\ x_n \end{pmatrix}, \qquad \vec{b} = \begin{pmatrix} b_1 \\ b_2 \\ \vdots \\ b_n \end{pmatrix}.$$

Our goal is to solve for $\vec{x}$. In this chapter we study direct solvers, which compute the exact solution in a finite number of steps. In the next chapter we will learn iterative solvers that generate approximate solutions through iterations.

Overview of topics in this chapter:

- Naive/Simple/Basic Gaussian elimination.
- Review on matrices: regularity and condition number.
- Different types of matrix A: banded, sparse, full.
- Direct solver for tri-diagonal systems.

117

6.2 Naive Gaussian Elimination: Simplest Version

Consider the system of linear equations in (6.1.1). The basic Gaussian elimination method requires two steps.

Step 1: We put the system into an upper triangular form, by a procedure called *forward elimination*. We denote by (i) the ith equation, and use the same notation a_{ij} for the current value of the coefficients, updated at each operation.

We begin by multiplying equation (1) by the factor $-\frac{a_{21}}{a_{11}}$ and add it to equation (2). We indicate this operation by means of the notation

$$(2) \leftarrow (2) - (1) \times \frac{a_{21}}{a_{11}}.$$

In this way, the first coefficient in the second row becomes zero: $a_{21} = 0$. Next, we multiply equation (1) by the factor $-\frac{a_{31}}{a_{11}}$ and add it to equation (3). Performing the operation

$$(3) \leftarrow (3) - (1) \times \frac{a_{31}}{a_{11}}$$

the first coefficient in the third row becomes zero: $a_{31} = 0$. Continuing in this way, the operations

$$(i) \leftarrow (i) - (1) \times \frac{a_{i1}}{a_{11}}$$

for $i = 2, 3, \cdots, n$ will lead to $a_{i1} = 0$ for $i = 2, 3, \cdots, n$. The system (6.1.1) becomes

$$\begin{cases} a_{11}x_1 + a_{12}x_2 + \cdots + a_{1n}x_n = b_1, & (1) \\ a_{22}x_2 + \cdots + a_{2n}x_n = b_2, & (2) \\ \qquad\qquad \vdots \\ a_{n2}x_2 + \cdots + a_{nn}x_n = b_n. & (n) \end{cases} \qquad (6.2.2)$$

Notice that all the entries in the first column, except a_{11}, are now zero.

Next, consider the second column. We begin by multiplying equation (2) by the factor $-\frac{a_{32}}{a_{22}}$ and add it to equation (3), i.e.,

$$(3) \leftarrow (3) - (2) \times \frac{a_{32}}{a_{22}}.$$

In this way, the second coefficient in the third row becomes zero: $a_{32} = 0$. Continuing in the same way, the operations

$$(i) \leftarrow (i) - (2) \times \frac{a_{i2}}{a_{22}}$$

for $i = 3, 4, \cdots, n$ will lead to $a_{i2} = 0$ for $i = 3, 4, \cdots, n$. The system (6.2.2) becomes

$$\begin{cases} a_{11}x_1 + a_{12}x_2 + a_{13}x_3 + \cdots + a_{1n}x_n = b_1, & (1) \\ a_{22}x_2 + a_{13}x_3 + \cdots + a_{2n}x_n = b_2, & (2) \\ a_{13}x_3 + \cdots + a_{3n}x_n = b_3, & (3) \\ \qquad\qquad \vdots \\ a_{13}x_3 + \cdots + a_{nn}x_n = b_n. & (n) \end{cases}$$

Notice that all the entries in the second column, except a_{12} and a_{22}, are now zero.

We can continue the process through all other columns, and make all the coefficients under the diagonal equal to zero, i.e., $a_{ij} = 0$ for $i > j$.

In summary, the algorithm involves 3 for-loops:

$$
\begin{aligned}
&\text{for } j = 1, 2, 3, \cdots, n - 1 \\
&\qquad \text{for } i = j + 1, j + 2, \cdots, n \\
&\qquad\qquad (i) \leftarrow (i) - (j) \times \frac{a_{ij}}{a_{jj}}, \\
&\qquad \text{end} \\
&\text{end}
\end{aligned}
$$

Note that there is an additional for-loop when we add $-(j) \times \frac{a_{ij}}{a_{jj}}$ onto equation (i).

Through this process, we transform the original system (6.1.1) into an equivalent system, having upper triangular form

$$
\begin{cases}
a_{11}x_1 + a_{12}x_2 + \cdots + a_{1n}x_n = b_1 & (1) \\
\qquad\quad a_{22}x_2 + \cdots + a_{2n}x_n = b_2 & (2) \\
\qquad\qquad\qquad\qquad\qquad \vdots & \\
\qquad\qquad\qquad\qquad a_{nn}x_n = b_n & (n)
\end{cases}
\tag{6.2.3}
$$

Note that here the coefficients a_{ij} and b_i are different from those in (6.1.1).

Step 2:   We then solve the system (6.2.3) through a *backward substitution*, which uses 2 for-loop, and obtain the exact solution. In other words, we first solve the last equation and find x_n. Then we insert the value of x_n in the previous equation, and find x_{n-1}, etc.

$$
\begin{aligned}
&x_n = \frac{b_n}{a_{nn}}, \\
&\text{for } i = n - 1, n - 2, \cdots, 1 \\
&\qquad x_i = \frac{1}{a_{ii}} \left(b_i - \sum_{j=i+1}^{n} a_{ij} x_j \right) \\
&\text{end}
\end{aligned}
$$

Note that the summation sign needs an additional for-loop.

Potential difficulty: In step 1, the algorithm requires a division by a_{kk}. If one of the diagonal elements a_{kk} is very close to or equal to 0, the computation breaks down.

Example 6.1. Solve the following 3×3 system:

$$\begin{cases} x_1 + x_2 + x_3 \quad\ = 1 & (1) \\ 2x_1 + 4x_2 + 4x_3 \ = 2 & (2) \\ 3x_1 + 11x_2 + 14x_3 = 6 & (3) \end{cases}$$

Answer. The forward elimination process goes as follows:

$$\begin{array}{lll} (1) * (-2) + (2): & 2x_2 + 2x_3 = 0 & (2') \\ (1) * (-3) + (3): & 8x_2 + 11x_3 = 3 & (3') \\ (2') * (-4) + (3'): & 3x_3 = 3 & (3'') \end{array}$$

In this way, we obtain the upper triangular system:

$$\begin{cases} x_1 + x_2 + x_3 = 1, \\ \quad\ 2x_2 + 2x_3 = 0, \\ \quad\quad\quad\ 3x_3 = 3. \end{cases}$$

The backward substitution process will now solve for the unknowns:

$$x_3 = 1,$$
$$x_2 = \frac{1}{2}(0 - 2x_3) = -1,$$
$$x_1 = 1 - x_2 - x_3 = 1.$$

Work amount: Assuming that A is an $n \times n$ matrix (corresponding to a system of n equations in n unknowns), we can count the number of arithmetic operations required by the algorithm. Here, one flop is one float number arithmetic operation, such as $+$, $-$, $*$, or $/$.

In the forward-elimination step, we need $\frac{1}{3}(n^3 - n)$ flops.

In the backward-substitution step, we need $\frac{1}{2}(n^2 - n)$ flops.

The total number of arithmetic operations is approximately $\frac{1}{3}n^3$.

This algorithm is <u>very slow</u> for large n. If $n = 10^3$, then it takes about 10^9 flops to solve the system. This method is unpractical for large scale computations. A more efficient algorithm is needed. Ideally, the entire computation should require between $\mathcal{O}(n)$ and $\mathcal{O}(n^2)$ flops, depending on the matrix A.

In Matlab, a smart version is available, with the scaled partial pivoting. Details are in the next section. Syntax to use it in Matlab is:

```
x=A\b;
x=inv(A)*b;   % both of them work
```

6.3 Gaussian Elimination with Pivoting (optional)

6.3.1 *Difficulties with Naive Gaussian Elimination*

First, we give an example where things go wrong with the naive Gaussian elimination.

Example 6.2. Solve the system with 3 significant digits.

$$\begin{cases} 0.001x_1 \ -x_2 = -1 & (1) \\ x_1 +2x_2 = 3 & (2) \end{cases}$$

Answer. Write it with 3 significant digits

$$\begin{cases} 1.00 \times 10^{-3}x_1 -1.00x_2 = -1.00 & (1) \\ 1.00x_1 +2.00x_2 = 3.00 & (2) \end{cases}$$

Now, $(1) * (-1.00 \times 10^3) + (2)$ gives

$$(1.00 \times 10^3 + 2.00)x_2 = 1.00 \times 10^3 + 3.00$$
$$\implies \quad 1.00 \times 10^3 x_2 = 1.00 \times 10^3 \quad \text{(We had to drop the last digit !)}$$
$$\implies \quad x_2 = 1.00.$$

Put this back into (1) and solve for x_1:

$$x_1 = \frac{1}{0.001}(-1.00 + 1.00x_2) = \frac{1}{0.001} \cdot 0.$$

Note that we did not get the solution at all! Here the answer for x_1 is wrong.

The trouble is that we have divided by 0.001, which is a very small number. This has caused a loss of significant digits.

To get around this difficulty, we change the order of two equations.

$$\begin{cases} 1.00x_1 +2.00x_2 = 3.00 & (1) \\ 1.00 \times 10^{-3}x_1 -1.00x_2 = -1.00 & (2) \end{cases}$$

Now run the whole procedure again: $(1) * (-0.001) + (2)$ will give us $x_2 = 1.00$. Inserting this value back in (1):

$$x_1 = 3.00 - 2.00x_2 = 1.00.$$

Our solution is now correct up to 3 digits.

Conclusion: The order in which equations are written can be important, to achieve an accurate computation!

6.3.2 *Gaussian Elimination with Partial Pivoting*

If we only exchange rows in the Gaussian elimination procedure, this leads to the *partial pivoting* algorithm. The underlying idea in this pivoting process is to permute rows so that, when we divide by the diagonal elements a_{ii}, we achieve the largest possible denominators.

Procedure:

(i) Find the smallest index k such that

$$|a_{k1}| \geq |a_{i1}|, \quad i = 1, \cdots, n.$$

Exchange equations (k) and (1), and do one step of the Gaussian elimination algorithm. You get

$$\begin{cases} a_{k1}x_1 + a_{k2}x_2 + \cdots + a_{kn}x_n = b_k & (1) \\ a_{22}x_2 + \cdots + a_{2n}x_n = b_2 & (2) \\ \quad \vdots \\ a_{12}x_2 + \cdots + a_{1n}x_n = b_1 & (k) \\ \quad \vdots \\ a_{n2}x_2 + \cdots + a_{nn}x_n = b_n & (n) \end{cases}$$

(ii) Repeat step (i) for the remaining $(n-1) \times (n-1)$ system, and so on.

This algorithm is used in Example 6.2, where we switched equations (1) and (2) before performing the Gaussian elimination.

6.3.3   *When Partial Pivoting is not Effective*

In some cases, the partial pivoting algorithm does not work well. Consider again the system in Example 6.2. Multiplying the first equation by 10^4, we obtain the equivalent system

$$\begin{cases} 10x_1 - 10^4 x_2 = -10^4 & (1) \\ x_1 + 2x_2 = 3 & (2) \end{cases}$$

Since $a_{11} > a_{21}$, the partial pivoting will not switch the equations for the Gaussian Elimination. Computing again with 2 significant digits, we multiply equation (1) with 0.1 and subtract it from equation (2), obtaining

$$(1.00^3 + 2.00)x_2 = 1.00 \times 10^3 + 3.00.$$

This yields the same result as in Example 6.2, i.e., $x_2 = 1.00$. The backward substitution now gives $x_1 = 0$, and we get an inaccurate solution.

To see what is causing this difficulty, we now do a more abstract analysis of the situation. Consider the system

$$\begin{cases} a_{11}x_1 + a_{12}x_2 = b_1, \\ a_{21}x_1 + a_{22}x_2 = b_2. \end{cases}$$

Assume that we have computed $\tilde{x}_2 = x_2 + \varepsilon_2$ where ε_2 is the error (machine error, roundoff error, etc.).

We then compute x_1 with this $\tilde{x}_2$:

$$\tilde{x}_1 = \frac{1}{a_{11}}(b_1 - a_{12}\tilde{x}_2)$$

$$= \frac{1}{a_{11}}(b_1 - a_{12}x_2 - a_{12}\varepsilon_2)$$

$$= \underbrace{\frac{1}{a_{11}}(b_1 - a_{12}x_2)}_{x_1} - \underbrace{\frac{a_{12}}{a_{11}}\varepsilon_2}_{\varepsilon_1}$$

$$= \quad x_1 \quad - \quad \varepsilon_1$$

Note that $\varepsilon_1 = \frac{a_{12}}{a_{11}}\varepsilon_2$. The error in x_2 is multiplied by the factor $\frac{a_{12}}{a_{11}}$.

Therefore, in order to get an accurate solution for x_1, we would like the factor $\frac{a_{12}}{a_{11}}$ to be as small as possible. Similarly, in order to get an accurate solution for x_2, we would like the factor $\frac{a_{21}}{a_{22}}$ to be as small as possible. It is not the absolute value of a_{11} that counts, but its relative size comparing to other entries in the same row. This basic idea leads to the *scaled partial pivoting* method.

6.3.4 *Gaussian Elimination with Scaled Partial Pivoting*

The main idea in this pivoting process is to use $\max_{k \leq i \leq n} |a_{ik}/s_i|$ for a_{ii} where s_i is a scaling factor for equation number i. To simplify the algorithm, the scaling factor is computed only once at the beginning of the algorithm, and is not updated later. This is a very effective algorithm and yet cheap to compute. Of course there are other pivoting algorithms, including a complete pivoting algorithm that even exchanges the columns of the coefficient matrix. We do not pursue them here.

The detailed algorithm is described below.

Procedure:

(1) Compute a scaling vector

$$\vec{s} = [s_1, s_2, \cdots, s_n], \quad \text{where} \quad s_i = \max_{1 \leq j \leq n} |a_{ij}|.$$

Keep this $\vec{s}$ for the rest of the computation.

(2) Find the index k such that

$$\left|\frac{a_{k1}}{s_k}\right| \geq \left|\frac{a_{i1}}{s_i}\right|, \quad i = 1, \cdots, n.$$

Exchange equations (k) and (1), and do one step of the Gaussian elimination algorithm. You get

$$\begin{cases} a_{k1}x_1 + a_{k2}x_2 + \cdots + a_{kn}x_n = b_k & (k) \\ \quad a_{22}x_2 + \cdots + a_{2n}x_n = b_2 & (2) \\ \qquad\qquad\qquad\vdots \\ \quad a_{12}x_2 + \cdots + a_{1n}x_n = b_1 & (1) \\ \qquad\qquad\qquad\vdots \\ \quad a_{n2}x_2 + \cdots + a_{nn}x_n = b_n & (n) \end{cases}$$

(3) Repeat (2) for the remaining $(n-1) \times (n-1)$ system, and continue through all columns. Note that s_i is unchanged, and is always associated with the original number i for the equation.

Example 6.3. Solve the system using scaled partial pivoting.

$$\begin{cases} x_1 + 2x_2 + x_3 = 3 & (1) \\ 3x_1 + 4x_2 + 0x_3 = 3 & (2) \\ 2x_1 + 10x_2 + 4x_3 = 10 & (3) \end{cases}$$

Answer. We follow the various steps.

(i) Compute $\vec{s}$.

$$\vec{s} = [2, 4, 10].$$

(ii) We have

$$\frac{a_{11}}{s_1} = \frac{1}{2}, \quad \frac{a_{21}}{s_2} = \frac{3}{4}, \quad \frac{a_{31}}{s_3} = \frac{2}{10}, \quad \implies \quad k = 2.$$

Exchange equations (1) and (2), and do one step of elimination

$$\begin{cases} 3x_1 + 4x_2 + 0x_3 = 3 & (2) \\ \frac{2}{3}x_2 + x_3 = 2 & (1') = (1) + (2) * \left(-\frac{1}{3}\right) \\ \frac{22}{3}x_2 + 4x_3 = 8 & (3') = (3) + (2) * \left(-\frac{2}{3}\right) \end{cases}$$

(iii) For the 2×2 system,

$$\frac{\bar{a}_{12}}{s_1} = \frac{2/3}{2} = \frac{1}{3}, \quad \frac{\bar{a}_{32}}{s_3} = \frac{22/3}{10} = \frac{22}{30} \quad \implies \quad k = 3.$$

Exchange $(3')$ and $(1')$ and do one step of elimination

$$\begin{cases} 3x_1 + 4x_2 + 0x_3 = 3 & (2) \\ \frac{22}{3}x_2 + 4x_3 = 8 & (3') \\ \frac{7}{11}x_3 = \frac{14}{11} & (1'') = (1') + (3') * \left(-\frac{1}{11}\right) \end{cases}$$

(iv) Backward substitution gives

$$x_3 = 2, \qquad x_2 = 0, \qquad x_1 = 1.$$

6.4 LU-Factorization (optional)

Here the idea is to write the coefficient matrix A as a product: $A = LU$, where

L: lower triangular matrix with unit diagonal,

U: upper triangular matrix.

We denote

$$
L = \begin{pmatrix} 1 & 0 & 0 & \cdots & 0 \\ l_{21} & 1 & 0 & \cdots & 0 \\ l_{31} & l_{32} & 1 & \cdots & 0 \\ \vdots & \vdots & & \ddots & \vdots \\ l_{n1} & l_{n2} & l_{n3} & \cdots & 1 \end{pmatrix}, \quad U = \begin{pmatrix} u_{11} & u_{12} & \cdots & u_{1,(n-1)} & u_{1n} \\ 0 & u_{22} & \cdots & u_{2,(n-1)} & u_{2n} \\ \vdots & & \ddots & \vdots & \vdots \\ 0 & 0 & \cdots & u_{(n-1),(n-1)} & u_{(n-1),n} \\ 0 & 0 & \cdots & 0 & u_{nn} \end{pmatrix}
$$

Theorem 6.1. *If $Ax = b$ can be solved by Gaussian elimination without pivoting, then we can write $A = LU$ uniquely.*

We can use the LU-factorization to solve $Ax = b$. Let $y = Ux$, then we have

$$
\begin{cases} U x = y, \\ L y = b. \end{cases}
$$

We need to solve two triangular systems, which are rather easy. We first solve y (by forward substitution), then solve x (by backward substitution). Each substitution takes about n^2-flops.

With pivoting, we have

$$
LU = PA
$$

where P is the pivoting matrix.

The LU-factorization is programmed in Matlab. One can use it as follows, to solve $Ax = b$:

```
> [L,U]=lu(A);
> y = L \ b;
> x = U \ y;
```

6.5   Matlab Simulations

Our Matlab function for naive Gaussian elimination looks like this:

```
function x = naiv_gauss(A,b);
n = length(b);   x = zeros(n,1);

% forward elimination
for k=1:n-1
  for i=k+1:n
    xmult = A(i,k)/A(k,k);
    for j=k+1:n
```

```
        A(i,j) = A(i,j)-xmult*A(k,j);
      end
      b(i) = b(i)-xmult*b(k);
   end
end

% back substitution
x(n) = b(n)/A(n,n);
for i=n-1:-1:1
  sum = b(i);
  for j=i+1:n
    sum = sum-A(i,j)*x(j);
  end
  x(i) = sum/A(i,i);
end
```

For example, we want to solve the system

$$\begin{pmatrix} 1^9 & 1^8 & 1^7 & \cdots & 1 & 1 \\ 2^9 & 2^8 & 2^7 & & 2 & 1 \\ 3^9 & 3^8 & 3^7 & & 3 & 1 \\ \vdots & & & & \vdots \\ 10^9 & 10^8 & 10^7 & \cdots & 10 & 1 \end{pmatrix} \begin{pmatrix} x_1 \\ x_2 \\ x_3 \\ \vdots \\ x_{10} \end{pmatrix} = \begin{pmatrix} 2 \\ 3 \\ 4 \\ \vdots \\ 11 \end{pmatrix}.$$

Note that the coefficient matrix here is a Vandermonde matrix. The exact solution is

$$(x_1, x_2, \ldots, x_{10}) = (0, 0, 0, \cdots, 0, 1, 1).$$

Comparing our naive Gaussian Elimination code with the Matlab operation A\b, we have the following result:

```
>> naiv_gauss(A,b)          >> A\b
ans =                        ans =
    0.00000000000088            -0.00000000000000
   -0.00000000004331             0.00000000000000
    0.00000000091111            -0.00000000000000
   -0.00000001068735             0.00000000000000
    0.00000007662105            -0.00000000000004
   -0.00000034609131             0.00000000000017
    0.00000097786263            -0.00000000000049
   -0.00000165121338             0.00000000000086
    1.00000149290615             0.99999999999921
    0.99999945973353             1.00000000000029
```

```
maxerror=                        maxerror=
  1.6e-6                           8.6e-13
```

Recall that the Vandermonde matrix has very big condition number for large n. The round off errors are magnified by a huge factor in the solution. The max error using the naive Gaussian Elimination is much larger than the one using pivoting.

LU factorization is coded in Matlab, and can be used in the following way to solve systems of linear equations.

```
>> % make a randon 4x4 matrix
>> A = rand(4)
A =
    0.3961    0.0850    0.6639    0.1191
    0.5327    0.0981    0.0208    0.3344
    0.7264    0.4951    0.3609    0.1855
    0.3239    0.8650    0.0558    0.6908

>> x = [1;1;1;1];    % would be the solution
>> b = A*x;       % make a rhs st the given x
>> [L,U] = lu(A)  % LU factorization
L =
    0.5453   -0.2872    1.0000         0
    0.7334   -0.4114   -0.6573    1.0000
    1.0000         0         0         0
    0.4460    1.0000         0         0
U =
    0.7264    0.4951    0.3609    0.1855
         0    0.6442   -0.1052    0.6081
         0         0    0.4369    0.1927
         0         0         0    0.5752

>> % back substitution and solving
>> v = L\b;
>> xs = U\v
xs =
    1.0000
    1.0000
    1.0000
    1.0000
```

6.6 Tridiagonal and Banded Systems

Tridiagonal systems are systems of linear equations where the coefficients matrix A is a tridiagonal, with non-zero elements only on the diagonal, the upper diagonal and the lower diagonal.

We often write

$$A = \text{tridiag}(a_i, d_i, c_i),$$

which means

$$A = \begin{pmatrix} d_1 & c_1 & 0 & \cdots & 0 & 0 & 0 \\ a_1 & d_2 & c_2 & \cdots & 0 & 0 & 0 \\ 0 & a_2 & d_3 & \cdots & 0 & 0 & 0 \\ \vdots & & \ddots & \ddots & \ddots & & \vdots \\ 0 & 0 & 0 & \cdots & d_{n-2} & c_{n-2} & 0 \\ 0 & 0 & 0 & \cdots & a_{n-2} & d_{n-1} & c_{n-1} \\ 0 & 0 & 0 & \cdots & 0 & a_{n-1} & d_n \end{pmatrix}.$$

A system $Ax = b$ where A is a tridiagonal matrix can be solved very efficiently using Gaussian elimination, provided that the system is non-singular (for example, diagonal dominant):

- Forward elimination (without pivoting). This procedure now takes only one for-loop.

> for $i = 2, 3, \cdots, n$
>
> $\quad d_i \leftarrow d_i - \frac{a_{i-1}}{d_{i-1}} c_{i-1}$
>
> $\quad b_i \leftarrow b_i - \frac{a_{i-1}}{d_{i-1}} b_{i-1}$
>
> end

At the end of this step, the A matrix looks like

$$A = \begin{pmatrix} d_1 & c_1 & 0 & \cdots & 0 & 0 & 0 \\ 0 & d_2 & c_2 & \cdots & 0 & 0 & 0 \\ 0 & 0 & d_3 & \cdots & 0 & 0 & 0 \\ \vdots & & \ddots & \ddots & \ddots & & \vdots \\ 0 & 0 & 0 & \cdots & d_{n-2} & c_{n-2} & 0 \\ 0 & 0 & 0 & \cdots & 0 & d_{n-1} & c_{n-1} \\ 0 & 0 & 0 & \cdots & 0 & 0 & d_n \end{pmatrix}$$

(with different coefficients c_i, d_i).

- Backward substitution. This procedure also requires only one for-loop

> $x_n \leftarrow b_n/d_n$
>
> for $i = n - 1, n - 2, \cdots, 1$
>
> $\quad x_i \leftarrow \frac{1}{d_i}(b_i - c_i x_{i+1})$
>
> end

Since each step has only one for-loop, the total number of flops for this algorithm is of order $\mathcal{O}(n)$.

Recall that the matrix for natural cubic splines is tridiagonal, and a simple Gaussian Elimination Matlab code was provided there. Review that code!

Penta-diagonal system. (Optional) A penta-diagonal matrix is when the matrix contains non-zeros element only in its diagonal, the 2 upper diagonals, and the 2 lower diagonals. We can write

$$A = \text{pentadiag}(e_i, a_i, d_i, c_i, f_i)$$

which means

$$A = \begin{pmatrix} d_1 & c_1 & f_1 & 0 & 0 & \cdots & 0 & 0 & 0 \\ a_1 & d_2 & c_2 & f_2 & 0 & \cdots & 0 & 0 & 0 \\ e_1 & a_2 & d_3 & c_3 & f_3 & \cdots & 0 & 0 & 0 \\ & \ddots & \ddots & \ddots & \ddots & \ddots & & \vdots & \vdots \\ & & \ddots & \ddots & \ddots & \ddots & \ddots \\ & & & \ddots & \ddots & \ddots & \ddots & \ddots \\ 0 & 0 & 0 & & \cdots & d_{n-2} & c_{n-2} & f_{n-2} \\ 0 & 0 & 0 & 0 & \cdots & a_{n-2} & d_{n-1} & c_{n-1} \\ 0 & 0 & 0 & 0 & 0 & \cdots & e_{n-2} & a_{n-1} & d_n \end{pmatrix}$$

Banded matrix: A banded matrix has non-zero elements only around the diagonal, within certain "distance" from the diagonal. This is a more general description. In detail, A matrix looks like:

$$A = \begin{pmatrix} d_1 & \ddots & \ddots & \ddots & & & 0 \\ \ddots & d_2 & \ddots & \ddots & \ddots \\ \ddots & \ddots & d_3 & \ddots & \ddots & \ddots \\ \ddots & \ddots & \ddots & d_4 & \ddots & \ddots & \ddots \\ & \ddots & \ddots & \ddots & \ddots & \ddots & \ddots \\ & & \ddots & \ddots & \ddots & \ddots & \ddots \\ 0 & & & \ddots & \ddots & \ddots & d_n \end{pmatrix}$$

$$| \leftarrow k \rightarrow |$$

Here $k \geq 0$ is the *band width*, meaning

$$a_{ij} = 0 \qquad \text{for all} \quad |i - j| > k.$$

We have the special cases, when band width is small:

- Diagonal matrix: $k = 0$,
- Tri-diagonal matrix: $k = 1$,
- Penta-diagonal matrix: $k = 2$.

Gaussian elimination is efficient if $k << n$, i.e., k is much smaller than n.

6.7 Review of Linear Algebra

In this discussion we consider an $n \times n$ square matrix $A = \{a_{ij}\}$, and denote x a column vector of length n.

Diagonally dominant system. If

$$|a_{ii}| \geq \sum_{j=1, j \neq i}^{n} |a_{ij}|, \quad i = 1, 2, \cdots, n$$

then A is called *diagonally dominant*. If the $\geq$ sign is replaced by $>$, we said that A is *strictly diagonally dominant*. A strictly diagonally dominant matrix A has the following properties:

- A is non-singular, regular, invertible, A^{-1} exists, and $Ax = b$ has a unique solution.
- $Ax = b$ can be solved by Gaussian Elimination without pivoting.

One such example we have already seen is the system from natural cubic spline.

Vector and matrix norms. A norm of a vector x or a matrix A is a quantity that measures the "size" of the vector or matrix. It is usually denoted as

$$\|x\| \quad \text{and} \quad \|A\|.$$

There are various ways of defining a norm for the same vector or matrix, and we would use different subscripts to denote them, see below.

Basic properties of a norm: $x \in I\!\!R^n$ or $x \in I\!\!R^{n \times n}$. Then, $\|x\|$ satisfies

(1) $\|x\| \geq 0$, with $\|x\| = 0$ if and only if $x = 0$;
(2) $\|ax\| = |a| \cdot \|x\|$, for any constant a;
(3) $\|x + y\| \leq \|x\| + \|y\|$ (triangle inequality).

Examples of norms for a vector $x \in I\!\!R^n$. These are the mostly commonly used vector norms:

- $\|x\|_1 = \sum_{i=1}^{n} |x_i|,$ l_1-norm

- $\|x\|_2 = \left(\sum_{i=1}^{n} x_i^2 \right)^{1/2},$ l_2-norm

- $\|x\|_\infty = \max_{1 \leq i \leq n} |x_i|,$ l_∞-norm

Matrix norms. Now consider an $n \times n$ matrix A. Its norm $\|A\|$ can be defined by setting

$$\|A\| = \max_{x \neq 0} \frac{\|Ax\|}{\|x\|}.$$

Properties of matrix norm. Obviously we have

$$\|A\| \geq \frac{\|Ax\|}{\|x\|} \quad \Longrightarrow \quad \|Ax\| \leq \|A\| \cdot \|x\|.$$

In addition, calling I the identity matrix, we have

$$\|I\| = 1, \qquad \|AB\| \leq \|A\| \cdot \|B\|.$$

Examples of matrix norms. In the above formula, we can use any of the vector norms $\|\cdot\|_1, \|\cdot\|_2, \|\cdot\|_\infty$. The corresponding matrix norms are

$$l_1 \text{ - norm}: \qquad \|A\|_1 = \max_{1 \leq j \leq n} \sum_{i=1}^{n} |a_{ij}|,$$

$$l_2 \text{ - norm}: \qquad \|A\|_2 = \max_i \sqrt{|\lambda_i|}, \qquad \lambda_i : \text{ eigenvalues of } A^t A,$$

$$l_\infty \text{ - norm}: \qquad \|A\|_\infty = \max_{1 \leq i \leq n} \sum_{j=1}^{n} |a_{ij}|.$$

Here A^t denotes the transpose of the matrix A. In the special case where A is symmetric, i.e., $A = A^t$, we have the simpler formula

$$\|A\|_2 = \max_i |\lambda_i|, \qquad \lambda_i : \text{ eigenvalues of } A. \tag{6.7.4}$$

Eigen-pairs of A. If we can find a number λ and a nonzero vector v such that

$$Av = \lambda v,$$

then λ is called an **eigenvalue** of A and v is the corresponding **eigenvector**. The couple (λ, v) is often referred to as an eigen-pair for A. We observe that

$$(A - \lambda I)v = 0 \quad \Longrightarrow \quad \det(A - \lambda I) = 0.$$

The eigenvalues of A can thus be found by computing the roots of the polynomial

$$p(\lambda) = \det(A - \lambda I).$$

If A is an $n \times n$ matrix, then $p(\lambda)$ is a polynomial of degree n. Therefore A has n eigenvalues, counting multiplicity.

Properties of eigenvalues, eigenvectors, and matrix norm: Assume that A is an invertible, symmetric matrix, and call A^{-1} its inverse. Then the eigenvalues of A^{-1} satisfy

$$\lambda_i(A^{-1}) = \frac{1}{\lambda_i(A)}.$$

By (6.7.4), this implies

$$\|A^{-1}\|_2 = \max_i |\lambda_i(A^{-1})| = \max_i \frac{1}{|\lambda_i(A)|} = \frac{1}{\min_i |\lambda_i(A)|}.$$

Condition number of a matrix. We now discuss the condition number of a matrix A and its significance.

Consider the system $Ax = b$, which we want to solve numerically. We want to understand how sensitive is the solution with respect to some small perturbation on the load vector b. A motivation for this could be that there is always the roundoff error in representing the load vector. We denote this perturbation as a vector p, with small entries, and consider the perturbed system:

$$A\bar{x} = b + p.$$

Here $\bar{x}$ is the solution of the perturbed system, which will be slightly different from x. We want to find a way of measuring the difference between x and $\bar{x}$, and how it relates to the change in the right-hand side.

The relative error in the perturbation of b is

$$e_b = \frac{\|p\|}{\|b\|}.$$

The relative change in solution x is

$$e_x = \frac{\|\bar{x} - x\|}{\|x\|}.$$

We wish to find a relation between e_x and e_b. We have

$$A(\bar{x} - x) = p,$$

which implies

$$\bar{x} - x = A^{-1}p.$$

Therefore

$$e_x = \frac{\|\bar{x} - x\|}{\|x\|} = \frac{\|A^{-1}p\|}{\|x\|} \leq \frac{\|A^{-1}\| \cdot \|p\|}{\|x\|}.$$

Since $Ax = b$, we have

$$\|b\| = \|Ax\| \leq \|A\| \cdot \|x\|.$$

Therefore

$$\frac{1}{\|x\|} \leq \frac{\|A\|}{\|b\|}.$$

Putting these together, we get

$$e_x \leq \|A^{-1}\| \cdot \|p\| \cdot \frac{\|A\|}{\|b\|} = \|A\| \cdot \|A^{-1}\| e_b \doteq \kappa(A)e_b.$$

Here the constant

$$\kappa(A) \doteq \|A\| \cdot \|A^{-1}\|$$

is called *the condition number* of A. We see that the error in b propagates with a factor of $\kappa(A)$ into the solution.

If $\kappa(A)$ is very large, the system $Ax = b$ is very sensitive to perturbations, therefore difficult to solve. Moreover, a small error in the measurement of the load vector b can produce a large error in the solution x. In this case we say that the system is *ill-conditioned*.

When the matrix A is symmetric, using the l_2 matrix norm, the condition number becomes

$$\kappa(A) = \frac{\max_i |\lambda_i|}{\min_i |\lambda_i|}.$$

This is one reason why eigenvalues of A are very important.

Some useful Matlab commands related to this discussion include:

```
[L,U] = lu(A);    % LU-factorization (optional)
norm(x);          % vector norm
eig(A);           % eigenvalue/eigen vector of a matrix
cond(A);          % condition number of A
```

6.8   Homework Problems for Chapter 6

1. Gaussian Elimination

Consider the following 3×3 system.

$$
\begin{aligned}
3.3330\,x_1 + 15920\,x_2 - 10.333\,x_3 &= 7973.6, \\
2.2220\,x_1 + 16.710\,x_2 + 9.6120\,x_3 &= 0.96500, \\
-1.5611\,x_1 + 5.1792\,x_2 - 1.6855\,x_3 &= 2.7140.
\end{aligned}
$$

Verify that the solution is $x_1 = 1, x_2 = 0.5, x_3 = -1$.

Solve the following system with naive Gaussian elimination, by using a calculator (not Matlab). All calculations should use 5 significant digits. Show details of your work. What do you get? Any comments?

2. Gaussian Elimination in Matlab

In this problem we test the performance of the naive Gaussian elimination procedure and compare it with the linear system solver implemented in Matlab (which uses scaled partial pivoting). As a test problem, we solve the system

$$
\mathbf{Ax = b}
$$

where A is the *Vandermonde*-matrix:

$$
\mathbf{A} =
\begin{pmatrix}
c_0^n & c_0^{n-1} & \cdots & c_0 & 1 \\
c_1^n & c_1^{n-1} & \cdots & c_1 & 1 \\
\vdots & \vdots & & \vdots & \vdots \\
c_n^n & c_n^{n-1} & \cdots & c_n & 1
\end{pmatrix}
$$

where $\mathbf{c} = [c_0, c_1, \ldots, c_n]$ is a given vector. Since the main point of this problem is to compare the two algorithms, we choose a vector $\mathbf{b}$ which gives a known solution $\mathbf{x}$. For example, if the solution is $\mathbf{x} = \text{ones}(n) \doteq [1, 1, \cdots, 1]$, then the load vector is $\mathbf{b} = \mathbf{A} \cdot \text{ones}(n)$.

Solve the above system with two different algorithms:

- The naive Gaussian elimination without pivoting, using the function `naiv_gauss(A,b)`, which you can find in Section 6.5.
- Gaussian elimination with pivoting in Matlab, i.e., using the command `x = A\b;`

You should test it for the following 3 cases:

i) $\mathbf{c} = [0.2, 0.4, 0.6, 0.8, 1]^T$,
ii) $\mathbf{c} = [0.1, 0.2, 0.3, \ldots, 0.9, 1]^T$,
iii) $\mathbf{c} = [0.05, 0.10, 0.15, \ldots, 0.90, 0.95, 1]^T$.

Compare the two solutions. What do you observe? What do you think is happening? Comment on your results.

Hint! Construct your $\mathbf{A}$ and $\mathbf{b}$ in Matlab in the following way:

```
c = [0.2:0.2:1];
A = vander(c);      % make Van der Monde-matrix.
xsol = ones(c');    % make the column vector xsol=[1,1,1,1,1]'.
b = A*xsol;         % make the load vector b
```

What to hand in: Your script file, your results for all 3 cases, and your comments.

3. Application of system of linear equations

Figure 6.1 shows a mechanical structure which consists of 17 rigid sticks:

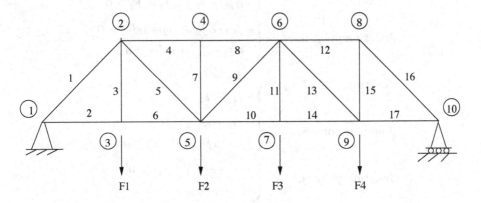

Fig. 6.1   Construction

The sticks form 3 squares, and all the triangles in the structure are isosceles right triangles. Therefore all the angles in the graph are 45° or 90°. Each point of junction is labelled with a circled number, and there are 10 such junctions. Note that the junction point no. 1 is fixed on the ground, while the junction point no. 10 is supported on wheels on a horizontal ground, so that it can move freely in the horizontal direction.

The forces F_1, F_2, F_3, F_4 are applied to the points no. 3, 5, 7, 9, respectively.

We assume that the structure is in equilibrium. This means, at each junction, the total force is zero, i.e., the total force components in x- and y-directions are both zero.

We wish to compute the force on each stick, for different values of the source loads F_1, F_2, F_3, F_4.

To simplify notation, we also write $\alpha = \sin(45°) = \cos(45°)$, since this appears in many equations.

We denote f_i as the force in the stick number i. At each junction, the two equations, representing the equilibrium of forces in the x and y direction, are listed below.

$$\alpha = \sin(45°) = \cos(45°)$$

Junction point 2:
$$\begin{cases} -\alpha f_1 + f_4 + \alpha f_5 = 0 \\ -\alpha f_1 + f_3 + \alpha f_5 = 0 \end{cases}$$

Junction point 3:
$$\begin{cases} -f_2 + f_6 = 0 \\ -f_3 + F_1 = 0 \end{cases}$$

Junction point 4:
$$\begin{cases} -f_4 + f_8 = 0 \\ f_7 = 0 \end{cases}$$

Junction point 5:
$$\begin{cases} -\alpha f_5 - f_6 + \alpha f_9 + f_{10} = 0 \\ -\alpha f_5 - f_7 + \alpha f_9 + F_2 = 0 \end{cases}$$

Junction point 6:
$$\begin{cases} -f_8 - \alpha f_9 + f_{12} + \alpha f_{13} = 0 \\ -\alpha f_9 + f_{11} + \alpha f_{13} = 0 \end{cases}$$

Junction point 7:
$$\begin{cases} -f_{10} + f_{14} = 0 \\ -f_{11} + F_3 = 0 \end{cases}$$

Junction point 8:
$$\begin{cases} -f_{12} + \alpha f_{16} = 0 \\ f_{15} - \alpha f_{16} = 0 \end{cases}$$

Junction point 9:
$$\begin{cases} -\alpha f_{13} - f_{14} + f_{17} = 0 \\ -\alpha f_{13} - f_{15} + F_4 = 0 \end{cases}$$

Junction point 10:
$$\begin{cases} -\alpha f_{16} - f_{17} = 0 \end{cases}$$

Be careful now! There are mistakes in some of these equations, left there on purpose to spice up the problem. Therefore, every student, with his or her mechanical talent, is challenged to double check each equation, and correct them in case of mistakes.

Once you are sure that you have a correct system, you need to find the forces in each stick when the whole construction is put under the following source forces:

a) $F_1 = 10$, $F_2 = 15$, $F_3 = 0$, $F_4 = 10$.
b) $F_1 = 15$, $F_2 = 0$, $F_3 = 0$, $F_4 = 10$.
c) $F_1 = 10$, $F_2 = 0$, $F_3 = 20$, $F_4 = 0$.
d) $F_1 = 0$, $F_2 = 10$, $F_3 = 10$, $F_4 = 0$.

You will end up in solving a system of linear equations, with can be written in matrix-vector form as $Ax = b$. Here the matrix A is sparse, i.e., most of its elements are 0. You are therefore encouraged to use the sparse matrix structure in Matlab. Check this out by 'help sparse' and 'help full'.

For example, "`sparse(m,n)`" generates a sparse matrix or vector of size $m \times n$, and "`full(V)`" makes a full matrix or vector out of a sparse one.

Even though our system is relatively small, and there is no urgent need to use the sparse matrix, nevertheless it is an important feature to learn.

What to hand on: Your script file, your results for a), b), c), d), and any comment that you would like to make.

Chapter 7

Fixed Point Iterative Solvers for Linear Systems

7.1 General Introduction to Iterative Solvers

Solutions of systems of linear equations play an important role in numerical simulation. In particular, the numerical solution of ordinary and partial differential equations is often reduced to solving a large system of linear equations. (These will be discussed in the last two chapters.) In these simulations, systems of linear equations can be very large. For example, the matrix A could be $10^6 \times 10^6$.

If the matrix A is large, solving the system $Ax = b$ using Gaussian elimination requires a long time. Faster methods are needed. Fortunately, in many problems, in particular in the discretization of differential equations, the coefficient matrix A has special properties which can be used to design very efficient algorithms. In this chapter we study the following problem:

Problem setting: Find an approximate solution to $Ax = b$, where $A \in I\!\!R^{n \times n}$ satisfies

- A is very large, for example $n = \mathcal{O}(10^6)$.
- A is sparse, with a large percentage of 0 entries.
- A is structured. For example, A is tri-diagonal, or has a specific block structure.

Numerical linear algebra is concerned with the computation of approximate solutions to $Ax = b$. Due to its importance, this is a very active research field, with an extensive literature. We mention two classes of iterative methods:

(1) Iterative solvers based on fixed-point iterations.

- Jacobi iterations,
- Gauss-Seidel iterations,
- SOR iterations.

(2) Krylov spaces techniques: More advanced and widely used, but not covered in these notes.

7.2   Jacobi Iterations

We want to solve the vector equation

$$Ax = b.$$

This can be written as a system of n equations:

$$\begin{cases} a_{11}x_1 + a_{12}x_2 + \cdots + a_{1n}x_n = b_1, \\ a_{21}x_1 + a_{22}x_2 + \cdots + a_{2n}x_n = b_2, \\ \qquad\qquad\qquad\qquad \vdots \\ a_{n1}x_1 + a_{n2}x_2 + \cdots + a_{nn}x_n = b_n. \end{cases} \qquad (7.2.1)$$

In order to compute approximate solutions by a fixed point iteration, we introduce some function G such that the solution to $Ax = b$ coincides with the solution of

$$x = G(x).$$

A simple way to do this is to choose $G(x) = Ax + x - b$. But there are better ways, leading to a faster convergence rate.

We begin by rewriting the system (7.2.1) in the following way. For the i-th equation, we keep the diagonal term x_i on the left and moving everything else to the right. This gives us

$$\begin{cases} x_1 = \frac{1}{a_{11}} \left(b_1 \qquad\qquad - a_{12}x_2 - \cdots - a_{1n}x_n \right), \\ x_2 = \frac{1}{a_{22}} \left(b_2 - a_{21}x_1 \qquad\quad - \cdots - a_{2n}x_n \right), \\ \quad \vdots \\ x_n = \frac{1}{a_{nn}} \left(b_2 - a_{n1}x_1 - a_{n2}x_2 - \cdots \qquad\qquad \right), \end{cases}$$

or in a compact form:

$$x_i = \frac{1}{a_{ii}} \left(b_i - \sum_{j=1,j\neq i}^{n} a_{ij}x_j \right), \qquad i = 1, 2, \cdots, n. \qquad (7.2.2)$$

This gives the **Jacobi iterations**:

- As initial guess, choose some vector $x^0 = (x_1^0, x_2^0, \cdots, x_n^0)$.
- For $k = 0, 1, 2, \ldots$, as soon as the vector $x^k = (x_1^k, \cdots, x_n^k)$ is found, the next approximation $x^{k+1} = (x_1^{k+1}, \cdots, x_n^{k+1})$ is computed by

$$x_i^{k+1} = \frac{1}{a_{ii}} \left(b_i - \sum_{j=1,j\neq i}^{n} a_{ij}x_j^k \right) \qquad \text{for } i = 1, \ldots, n.$$

Choices for the starting vector x^0. As initial guess x^0, one may choose any of the following:

- The vector $x^0 = (1, 1, \cdots, 1)$, with all entries $= 1$;
- The load vector: $x^0 = b$;
- If A is diagonally dominant, the best choice is:
$$x_i = b_i/a_{ii}, \qquad \text{for} \qquad i = 1, 2, \cdots, n.$$

Stop Criteria. The iteration can be continued for $k = 1, 2, 3, \cdots$ until any of the following criteria is met:

- x^k is close enough to x^{k-1}, for example $\|x^k - x^{k-1}\| \leq \varepsilon$ for some vector norm.
- The residual $r^k = Ax^k - b$ is small: for example $\|r^k\| \leq \varepsilon$.
- A maximum number of iterations, fixed in advance, is reached.

Remarks on the algorithm:

- At each iteration one has two vectors: x^k and x^{k+1}. This requires more memory space.
- At each step, the n components x_i^{k+1}, $i = 1, \ldots, n$ can be computed independently of each other. It is thus possible to do parallel computing.

We now look at how the iteration performs through an example.

Example 7.1. Compute an approximate solution to the following system using Jacobi iterations.
$$\begin{cases} 2x_1 - x_2 & = 0, \\ -x_1 + 2x_2 - x_3 = 1, \\ -x_2 + 2x_3 = 2. \end{cases}$$
Compare the result with the exact solution $(x_1, x_2, x_3) = (1, 2, 2)$.

Answer. Choose x^0 by setting $x_i^0 = b_i/a_{ii}$, so that
$$x^0 = (0, \ 0.5, \ 1).$$

The Jacobi iteration at step k is
$$\begin{cases} x_1^{k+1} = \frac{1}{2}x_2^k, \\ x_2^{k+1} = \frac{1}{2}(1 + x_1^k + x_3^k), \\ x_3^{k+1} = \frac{1}{2}(2 + x_2^k). \end{cases}$$
We perform a couple of iterations, and get
$$x^1 = (0.25, \ 1, \ 1.25),$$
$$x^2 = (0.5, \ 1.25, \ 1.5),$$
$$x^3 = (0.625, \ 1.5, \ 1.625).$$

Observations:

- Looks like it is converging. One should run a few more steps to be sure.
- In any case, the convergence is rather slow.

7.3   Gauss-Seidel Iterations

We consider now an improved version of the Jacobi algorithm. Observe that in the Jacobi iteration, we can write

$$
x_i^{k+1} = \frac{1}{a_{ii}} \left(b_i - \sum_{j=1,j\neq i}^{n} a_{ij} x_j^k \right)
$$

$$
= \frac{1}{a_{ii}} \left(b_i - \sum_{j=1}^{i-1} a_{ij} x_j^k - \sum_{j=i+1}^{n} a_{ij} x_j^k \right).
$$

If the computation is done in a sequential way, for $i = 1, 2, \cdots$, then in the first summation term, all x_j^k are already computed for step $k+1$. Assuming the iteration converges, then the newly computed values are better.

This idea leads to the **Gauss-Seidel iteration:**

As soon as the vector $x^k = (x_1^k, \dots, x_n^k)$ is found, the next approximation $x^{k+1} = (x_1^{k+1}, \dots, x_n^{k+1})$ is computed by

$$
x_i^{k+1} = \frac{1}{a_{ii}} \left(b_i - \sum_{j=1}^{i-1} a_{ij} x_j^{k+1} - \sum_{j=i+1}^{n} a_{ij} x_j^k \right), \qquad i = 1, 2, \dots, n.
$$

Remarks on the algorithm:

- It needs only one vector for both x^k and x^{k+1}, and we can simply over-write the values in x as we go through the algorithm. This means, the computation in the inner for-loop in the code simply becomes

$$
x_i = \frac{1}{a_{ii}} \left(b_i - \sum_{j=1}^{i-1} a_{ij} x_j - \sum_{j=i+1}^{n} a_{ij} x_j \right),
$$

where x is used for both x^k and x^{k+1}. This saves some memory space.
- The algorithm is sequential, hence it is not good for parallel computing.

We now go through an example using Gauss-Seidel iterations.

Example 7.2. Consider the same system as in Example 7.1. Perform Gauss-Seidel iterations, with initial guess $x^0 = (0, 0.5, 1)$. The iteration at step k is:

$$
\begin{cases}
x_1^{k+1} = \frac{1}{2} x_2^k, \\[2mm]
x_2^{k+1} = \frac{1}{2}(1 + x_1^{k+1} + x_3^k), \\[2mm]
x_3^{k+1} = \frac{1}{2}(2 + x_2^{k+1}).
\end{cases}
$$

We perform a couple of iterations, and get

$$
x^1 = (0.25,\ 1.125,\ 1.5625),
$$
$$
x^2 = (0.5625,\ 1.5625,\ 1.7813).
$$

Observation: It converges a bit faster than Jacobi iterations. It looks like the last component x_3^k converges faster than the first component x_1^k, because it uses the latest values of the other coordinates.

This suggests another idea, to improve the overall convergence. We could alternate the order of the variables in the inner for-loop. At one step we could sweep from $i = 1$ to $i = n$, and the next step sweep from $i = n$ to $i = 1$. This would even out a bit the convergence among components.

7.4   SOR Iterations

SOR (Successive Over Relaxation) is a more general iterative method. The basic idea can be applied to either Jacobi or Gauss-Seidel iterations. We consider a version that is based on Gauss-Seidel.

The iterations are performed with a parameter $w > 0$, which is called the *relaxation parameter*. The iteration at step k looks like:

$$x_i^{k+1} = (1 - w)x_i^k + w \cdot \frac{1}{a_{ii}} \left(b_i - \sum_{j=1}^{i-1} a_{ij}x_j^{k+1} - \sum_{j=i+1}^{n} a_{ij}x_j^k \right).$$

Note the second term is the Gauss-Seidel iteration multiplied by w. If $w \in [0, 1]$, then $w \geq 0$ and $(1 - w) \geq 0$. Therefore, the above right-hand side is some weighted average of the previous value x_i^k and the new value computed by the Gauss-Seidel iteration. Since we believe that the Gauss-Seidel result is the correct direction to go, we attach to it a higher weight. Usually, one chooses $1 < w < 2$. It can be shown that these iterations converge when $0 < w < 2$. They are classified as follows:

- $w = 1$: standard Gauss-Seidel;
- $0 < w < 1$: under relaxation;
- $1 < w < 2$: over relaxation. This is the range commonly used for SOR iterations.

We now give an example using SOR iterations.

Example 7.3. Perform SOR iterations on the same problem as in Example 7.1 with $w = 1.2$. The iteration at step k is:

$$\begin{cases} x_1^{k+1} = -0.2x_1^k + 0.6x_2^k, \\ x_2^{k+1} = -0.2x_2^k + 0.6*(1 + x_1^{k+1} + x_3^k), \\ x_3^{k+1} = -0.2x_3^k + 0.6*(2 + x_2^{k+1}). \end{cases}$$

Choosing $x^0 = (0, \ 0.5, \ 1)$ as our initial guess, we obtain

$$x^1 = (0.3, 1.28, 1.708),$$
$$x_2 = (0.708, 1.8290, 1.9442).$$

Recall that the exact solution is $(x_1, x_2, x_3) = (1, 2, 2)$. Observe that here the convergence is faster than with Jacobi or Gauss-Seidel iterations.

7.5 Matlab Simulations

We wish to solve a system:

$$Ax = b$$

where A is a 6×6 matrix

```
A =
    4    -1    -1     0     0     0
   -1     4     0    -1     0     0
   -1     0     4    -1    -1     0
    0    -1    -1     4     0    -1
    0     0    -1     0     4    -1
    0     0     0    -1    -1     4
```

and the load vector is:
    b=[1; 5; 0; 3; 1; 5].
Our approximations will then be compared with the exact solution:
    x=[1; 2; 1; 2; 1; 2].

We solve the system with iterative methods, choosing the initial value:

$$x^{(0)} = [0.25; 1.25; 0; 0.75; 0.25; 1.25].$$

Below are the results obtained with our three methods.

Jacobi iterations:

k	x1	x2	x3	x4	x5	x6
1	0.2500	1.2500	0	0.7500	0.2500	1.2500
2	0.5625	1.5000	0.3125	1.3750	0.5625	1.5000
3	0.7031	1.7344	0.6250	1.5781	0.7031	1.7344
4	0.8398	1.8203	0.7461	1.7734	0.8398	1.8203
5	0.8916	1.9033	0.8633	1.8467	0.8916	1.9033
6	0.9417	1.9346	0.9075	1.9175	0.9417	1.9346

7	0.9605	1.9648	0.9502	1.9442	0.9605	1.9648
8	0.9787	1.9762	0.9663	1.9699	0.9787	1.9762
9	0.9856	1.9872	0.9819	1.9797	0.9856	1.9872
10	0.9923	1.9913	0.9877	1.9890	0.9923	1.9913
11	0.9948	1.9953	0.9934	1.9926	0.9948	1.9953
12	0.9972	1.9968	0.9955	1.9960	0.9972	1.9968
13	0.9981	1.9983	0.9976	1.9973	0.9981	1.9983
14	0.9990	1.9988	0.9984	1.9985	0.9990	1.9988
15	0.9993	1.9994	0.9991	1.9990	0.9993	1.9994
16	0.9996	1.9996	0.9994	1.9995	0.9996	1.9996
17	0.9997	1.9998	0.9997	1.9996	0.9997	1.9998
18	0.9999	1.9998	0.9998	1.9998	0.9999	1.9998
19	0.9999	1.9999	0.9999	1.9999	0.9999	1.9999
20	1.0000	1.9999	0.9999	1.9999	1.0000	1.9999
21	1.0000	2.0000	1.0000	2.0000	1.0000	2.0000

It takes 21 iterations to reach the error tolerance of 10^{-4}.

Gauss-Seidel iterations:

k	x1	x2	x3	x4	x5	x6
1	0.2500	1.2500	0	0.7500	0.2500	1.2500
2	0.5625	1.5781	0.3906	1.5547	0.6602	1.8037
3	0.7422	1.8242	0.7393	1.8418	0.8857	1.9319
4	0.8909	1.9332	0.9046	1.9424	0.9591	1.9754
5	0.9594	1.9755	0.9652	1.9790	0.9852	1.9910
6	0.9852	1.9911	0.9873	1.9924	0.9946	1.9967
7	0.9946	1.9967	0.9954	1.9972	0.9980	1.9988
8	0.9980	1.9988	0.9983	1.9990	0.9993	1.9996
9	0.9993	1.9996	0.9994	1.9996	0.9997	1.9998
10	0.9997	1.9998	0.9998	1.9999	0.9999	1.9999
11	0.9999	1.9999	0.9999	2.0000	1.0000	2.0000
12	1.0000	2.0000	1.0000	2.0000	1.0000	2.0000
===	========	========	========	========	========	=======
13	1.0000	2.0000	1.0000	2.0000	1.0000	2.0000
14	1.0000	2.0000	1.0000	2.0000	1.0000	2.0000
15	1.0000	2.0000	1.0000	2.0000	1.0000	2.0000

We see that after 12 iterations the error tolerance 10^{-4} is reached.

SOR iterations with $w = 1.12$:

k	x1	x2	x3	x4	x5	x6
1	0.2500	1.2500	0	0.7500	0.2500	1.2500
2	0.6000	1.6280	0.4480	1.6813	0.7254	1.9239
3	0.7893	1.8964	0.8411	1.9434	0.9671	1.9841
4	0.9518	1.9831	0.9805	1.9922	0.9940	1.9980
5	0.9956	1.9986	0.9972	1.9992	0.9994	1.9998
6	0.9994	1.9998	0.9998	1.9999	1.0000	2.0000
7	0.9999	2.0000	1.0000	2.0000	1.0000	2.0000
8	1.0000	2.0000	1.0000	2.0000	1.0000	2.0000
=====	=====	=====	=====	=====	=====	=====
9	1.0000	2.0000	1.0000	2.0000	1.0000	2.0000
10	1.0000	2.0000	1.0000	2.0000	1.0000	2.0000

We see that after 8 iterations the error tolerance 10^{-4} is reached.

Errors Comparisons. The plots of the errors, for various numbers of iterations of the Jacobi, Gauss-Seidel and SOR algorithms, are given in Figure 7.1. It is clear that with SOR the error decreases more quickly than with the other methods.

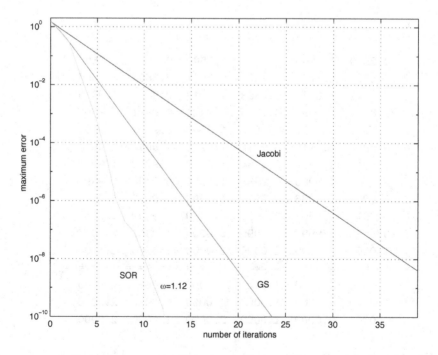

Fig. 7.1   Plots of error against number of iterations for Jacobi, Gauss-Sidel and SOR iterations.

7.6 Writing All Three Methods in Matrix-Vector Form

We want to solve $Ax = b$. We change it into a fixed-point problem, $x = G(x)$, where $G(x)$ is a vector valued function, which is linear in x. Then, we can write it as
$$x = G(x) = y + Mx$$
for some vector y and some matrix M.

Consider the splitting of the matrix A:
$$A = L + D + U$$
where

- L is the lower triangular part of A:
$$L = \{l_{ij}\}, \qquad l_{ij} = \begin{cases} a_{ij}, & i > j \\ 0 & i \le j \end{cases}$$

- D is the diagonal part of A:
$$D = \{d_{ij}\}, \qquad d_{ij} = \begin{cases} a_{ij} = a_{ii}, & i = j \\ 0 & i \ne j \end{cases}$$

- U is the upper triangular part of A:
$$U = \{u_{ij}\}, \qquad u_{ij} = \begin{cases} a_{ij}, & i < j \\ 0 & i \ge j \end{cases}$$

See the graph in Figure 7.2 for an illustration.

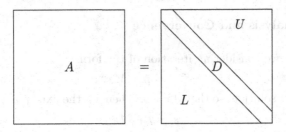

Fig. 7.2 Splitting of A.

We now have
$$Ax = (L + D + U)x = Lx + Dx + Ux = b.$$

Jacobi iterations: The Jacobi iterations can be written in a very compact form using the splitting of the matrix A. We have
$$Dx^{k+1} = b - Lx^k - Ux^k,$$
so
$$x^{k+1} = D^{-1}b - D^{-1}(L + U)x^k = y_J + M_J x^k,$$
where
$$y_J = D^{-1}b,$$
$$M_J = -D^{-1}(L + U).$$

Gauss-Seidel iterations: We can write
$$Dx^{k+1} + Lx^{k+1} = b - Ux^k,$$
so
$$x^{k+1} = (D+L)^{-1}b - (D+L)^{-1}Ux^k = y_{GS} + M_{GS}x^k,$$
where
$$y_{GS} = (D+L)^{-1}b,$$
$$M_{GS} = -(D+L)^{-1}U.$$

SOR iterations: We have
$$x^{k+1} = (1-w)x^k + wD^{-1}(b - Lx^{k+1} - Ux^k)$$
$$\implies \quad Dx^{k+1} = (1-w)Dx^k + wb - wLx^{k+1} - wUx^k$$
$$\implies \quad (D+wL)x^{k+1} = wb + [(1-w)D - wU]x^k,$$
so
$$x^{k+1} = (D+wL)^{-1}b + (D+wL)^{-1}[(1-w)D - wU]x^k = y_{SOR} + M_{SOR}x^k,$$
where
$$y_{SOR} = (D+wL)^{-1}b,$$
$$M_{SOR} = (D+wL)^{-1}[(1-w)D - wU].$$

7.7   Error Analysis and Convergence

To solve $Ax = b$, we consider an iteration of the form
$$x^{k+1} = y + Mx^k.$$
Let s be the exact solution, so that $As = b$. Then s is the fixed point of the iteration:
$$s = y + Ms.$$
At each iteration, we define the error vector:
$$e^k = x^k - s.$$
We have
$$e^{k+1} = x^{k+1} - s = y + Mx^k - (y + Ms) = M(x^k - s) = Me^k.$$
This yields a formula of how the error changes from one iteration to the next:
$$e^{k+1} = M e^k.$$
Taking the vector norm on both sides, we obtain
$$\|e^{k+1}\| = \|Me^k\| \le \|M\| \cdot \|e^k\|.$$
By induction, this implies
$$\|e^k\| \le \|M\|^k \|e^0\|, \qquad \text{where} \qquad e^0 = x^0 - s.$$

Theorem 7.1. *If $\|M\| < 1$ for some norm $\|\cdot\|$, then for any initial guess x^0 the iterations converge to the exact solution.*

We note that the convergence depends only on the iteration matrix M.

For our three methods, using the notation $A = D + L + U$, we have

- Jacobi: $M = -D^{-1}(L + U)$.
- G-S: $M = -(D + L)^{-1}U$.
- SOR: $M = (D + wL)^{-1}[(1 - w)D - wU]$.

In the first two cases the matrix M is uniquely determined by A. For the SOR method one can adjust the weight w in order to get a small value of $\|M\|$. For this reason, the SOR method is more flexible.

Example 7.4. Consider again the matrix

$$A = \begin{pmatrix} 2 & -1 & 0 \\ -1 & 2 & -1 \\ 0 & -1 & 2 \end{pmatrix}.$$

Its decomposition has the form

$$L = \begin{pmatrix} 0 & 0 & 0 \\ -1 & 0 & 0 \\ 0 & -1 & 0 \end{pmatrix}, \quad D = \begin{pmatrix} 2 & 0 & 0 \\ 0 & 2 & 0 \\ 0 & 0 & 2 \end{pmatrix}, \quad U = \begin{pmatrix} 0 & -1 & 0 \\ 0 & 0 & -1 \\ 0 & 0 & 0 \end{pmatrix}.$$

The corresponding matrices for the three methods are:

$$M_J = \begin{pmatrix} 0 & 0.5 & 0 \\ 0.5 & 0 & 0.5 \\ 0 & 0.5 & 0 \end{pmatrix},$$

$$M_{GS} = \begin{pmatrix} 0 & 0.5 & 0 \\ 0 & 0.25 & 0.5 \\ 0 & 0.125 & 0.25 \end{pmatrix},$$

$$M_{SOR} = \begin{pmatrix} -0.2 & 0.6 & 0 \\ -0.12 & 0.16 & 0.6 \\ -0.072 & 0.096 & 0.16 \end{pmatrix}, \quad \text{(with } w = 1.2\text{)}.$$

The various norms of these matrices are shown in the following table.

M	l_1 norm	l_2 **norm**	l_∞ norm
Jacobi	1	**0.707**	1
G-S	0.875	**0.5**	0.75
SOR	0.856	**0.2**	0.88

The l_2 norm is the most significant one. We see now why SOR converges fastest.

We now state a convergence theorem, without proof.

Theorem 7.2. *(Convergence of Iterative Methods) If A is diagonally dominant, i.e.,*

$$|a_{ii}| > \sum_{j=1,j\neq i}^{n} |a_{ij}|, \qquad \text{for all } i = 1, 2, \cdots, n.$$

Then all three iteration methods converge to the exact solution, for every initial choice of x^0.

Remarks:

- If A is not diagonally dominant, the iterations might still converge, but there is no guarantee that this will happen.
- If $\|M\| < 1$ in some norm, then the iteration converges in all of these vector norms. (Can you prove it?)

7.8 Homework Problems for Chapter 7

1. Comparing Various Methods

Solve the linear system with the four methods below (Do not use Matlab).

$$\begin{cases} 4\,x_1 + 3\,x_2 \quad\;\; = \quad 24 \\ 3\,x_1 + 4\,x_2 - \quad x_3 = \quad 30 \\ \quad\;\; -\;\; x_2 + 4\,x_3 = -24 \end{cases}$$

(a). Gaussian elimination (tridiagonal system).
(b). Jacobi's method.
(c). Gauss-Seidel's method.
(d). The SOR-method, with $\omega = 1.25$.

For the methods in (b), (c) and (d), write out the general iteration scheme, then do 2 iterations for each method, with initial guess

$$x^0 = [24/4,\, 30/4,\, -24/4] = [6,\, 7.5,\, -6].$$

Among the methods (b), (c), (d), which seems to work better for this example? Please comment.

2. SOR in Matlab

(a). Write a Matlab function which solves a system of linear equations $Ax = b$, with successive over relaxation (SOR) iterations. Assume here that A is a banded matrix with band width d, (so that $a_{ij} = 0$ for $|i - j| > d$). The inputs of the function are: A, b, a starting vector x_0, the band-width d, the relaxation parameter w, an error tolerance ε and the maximum number of iterations. The iteration stops when the error (you may use the residual $r = Ax - b$ measured in certain norm) is less than the tolerance, or when the maximum number of iterations is reached. The function should return the solution vector x and the number of iterations.

The first few lines in the function should look like this:

```
function [x,nit]=sor(A,b,x0,w,d,tol,nmax)
% SOR : solve linear system with SOR iteration
% Usage: [x,nit]=sor(A,b,x0,omega,d,tol,nmax)
% Inputs:
%         A : an n x n-matrix,
%         b : the rhs vector, with length n
%         x0 : the start vector for the iteration
%         tol: error tolerance
%         w: relaxation parameter, (1 < w < 2),
%         d : band width of A.
% Outputs::
%         x : the solution vector
```

%   nit: number of iterations

(b). Use your function sor to solve the following tridiagonal system, with

$$
A = \begin{pmatrix}
-2.011 & 1 & & & & & & & \\
1 & -2.012 & 1 & & & & & & \\
 & 1 & -2.013 & 1 & & & & & \\
 & & 1 & -2.014 & 1 & & & & \\
 & & & 1 & -2.015 & 1 & & & \\
 & & & & 1 & -2.016 & 1 & & \\
 & & & & & 1 & -2.017 & 1 & \\
 & & & & & & 1 & -2.018 & 1 \\
 & & & & & & & 1 & -2.019
\end{pmatrix},
$$

$$
b = \begin{pmatrix}
-0.994974 \\
1.57407 \cdot 10^{-3} \\
-8.96677 \cdot 10^{-4} \\
-2.71137 \cdot 10^{-3} \\
-4.07407 \cdot 10^{-3} \\
-5.11719 \cdot 10^{-3} \\
-5.92917 \cdot 10^{-3} \\
-6.57065 \cdot 10^{-3} \\
-0.507084
\end{pmatrix}, \qquad
x^0 = \begin{pmatrix}
0.95 \\
0.9 \\
0.85 \\
0.8 \\
0.75 \\
0.7 \\
0.65 \\
0.6 \\
0.55
\end{pmatrix}.
$$

Here b is the load vector and x^0 is the initial guess. Solve the system with an error tolerance $\varepsilon = 10^{-4}$, setting 100 as the maximum number of iterations. Try different values of w between 1 and 2 (this is called over-relaxation), for example $w = 1.0, 1.1, 1.2, \cdots, 1.9$. Search for the value of w that gives the fastest convergence (requiring the smallest number of iterations). Make a plot of number of iterations as a function of w.

What to hand in? Your Matlab file sor.m, the script, the running result, the plot, and whatever comments you have.

3. Jacobi Iterations in Matlab

Write a Matlab function, using Jacobi's method to solve a system of linear equations. Do it similarly as in Problem 2(a), but only for tri-diagonal systems.

Test it on the same system in Problem 2b), with error tolerance $\varepsilon = 10^{-4}$ and setting the maximum number of iterations $=100$. Compare the result with the result from SOR. Which method converges faster? Put your comments.

What to hand in? Your Matlab file jacobi.m, the script, the running result, and whatever comments you have.

Chapter 8

The Method of Least Squares

8.1 Problem Description

In this chapter we study the fundamental problem of data fitting. Let a data set be given

x	x_0	x_1	x_2	$\cdots$	x_m
y	y_0	y_1	y_2	$\cdots$	y_m

These data could come from observation (such as measured quantities) or from an experiment. Therefore it is quite possible that the values y_i will contain errors.

From modeling considerations, we expect that there should be some physical law relating y with x, i.e., $y = y(x)$ for a function $y(\cdot)$ of some special form. For example, we might guess that y has one of the following forms

- $y(x) = ax + b$,
- $y(x) = ax^2 + bx + c$,
- $y(x) = a\sin(x + b)$,
- $\cdots$

In these cases, we cannot hope to find the coefficients a, b, c so that all the equations $y_k = y(x_k)$ are satisfied exactly, for all $k = 0, 1, \ldots, m$. This would give us a set of $m + 1$ equations in only 2 or 3 unknowns.

Instead, we expect that our data will contain errors:

$$y_k = y(x_k) + e_k, \qquad k = 0, 1, \ldots, m, \qquad (8.1.1)$$

where e_k is error in the k-th measurement. Our goal is to determine these parameters a, b, c, so that the function $y(x)$ provides the "best" fit to the data, making these errors as small as possible.

Note that this is a very different problem from polynomial interpolation, which we learned in Chapter 2. In polynomial interpolation, given $m + 1$ data points we constructed a polynomial $p(x)$ of degree m which exactly interpolates all data: $p(x_k) = y_k$ for all $k = 0, 1, \ldots, m$. Here instead we restrict our search to polynomials of degree 1 (or 2), and try to determine the coefficients a, b (or a, b, c) in order to minimize the errors $e_0, e_1, \ldots, e_m$ in (8.1.1).

Overview. In this chapter we shall cover the following topics:

- Linear regression, with polynomials;
- Linear method of least squares, with parabolas and other functions;
- General linear method of least squares;
- Nonlinear method of least squares;
- Least squares approximations for continuous functions.

8.2 Linear Regression

Consider a set of data, say (x_k, y_k) for $k = 0, 1, \ldots, m$. By physical considerations, we expect that, if no errors were present, all these points should fit on a straight line $y = y(x) = ax + b$, for some choice of a, b. In practice however, because of experimental or measurement errors, the data will not lie exactly on a straight line. Our job is to find the straight line that best fits the data. This problem of determining the coefficients a, b in an optimal way is called *linear regression*. See Figure 8.1 for an illustration.

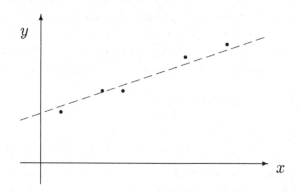

Fig. 8.1 Linear regression.

We can reformulate our problem as an optimization problem: Find a, b, such that, choosing $y(x) = ax + b$, the "error" becomes smallest possible.

Errors. How do we measure this error? At each data point we have an error

$$e_k = y(x_k) - y_k.$$

All these values form a vector: $e = (e_0, e_1, \ldots, e_m)$. One can then measure the total error by measuring the error vector e in some vector norm. The commonly used norms are:

(1) $e \doteq \max_{k} |e_k| = \max_{k} |y(x_k) - y_k|$ the l_∞ norm;

(2) $e \doteq \sum_{k=0}^{m} |e_k| = \sum_{k=0}^{m} |y(x_k) - y_k|$ the l_1 norm;

(3) $e \doteq \sum_{k=0}^{m} (e_k)^2 = \sum_{k=0}^{m} (y(x_k) - y_k)^2$ the l_2 norm.

Unfortunately, choices (1) and (2) lead to very difficult mathematical problems. The best choice is (3), which leads to the *method of least squares*. We thus define the error function

$$\psi(a, b) \doteq \sum_{k=0}^{m} (ax_k + b - y_k)^2. \tag{8.2.2}$$

Notice that $\psi(a, b)$ is the sum of the squares of all the errors, if our data set is approximated using the line $y(x) = ax + b$. Clearly, this depends on the values of the parameters a, b.

Minimization Problem. Our problem can now be stated as a minimization problem:

Find a and b such that the error function $\psi(a, b)$ defined in (8.2.2) is minimized.

How do we find a and b? Since ψ is a function of the two variables (a, b), at a point where it attains a minimum, its partial derivatives must vanish:

$$\frac{\partial \psi}{\partial a} = \frac{\partial \psi}{\partial b} = 0. \tag{8.2.3}$$

Recalling (8.2.2), from the above conditions we obtain

$$\frac{\partial \psi}{\partial a} = 0 \quad \Longrightarrow \quad \sum_{k=0}^{m} 2(ax_k + b - y_k)x_k = 0, \tag{8.2.4}$$

$$\frac{\partial \psi}{\partial b} = 0 \quad \Longrightarrow \quad \sum_{k=0}^{m} 2(ax_k + b - y_k) = 0. \tag{8.2.5}$$

This yields a system of two linear equations for the two unknowns a, b, which we write as

$$\begin{cases} \left(\sum_{k=0}^{m} x_k^2 \right) \cdot a + \left(\sum_{k=0}^{m} x_k \right) \cdot b = \sum_{k=0}^{m} x_k \cdot y_k, \\[2ex] \left(\sum_{k=0}^{m} x_k \right) \cdot a + (m+1) \cdot b = \sum_{k=0}^{m} y_k. \end{cases} \tag{8.2.6}$$

These are called the *normal equations*. Notice that, in order to write the normal equations, we need to compute the sums

$$\sum_{k=0}^{m} x_k^2, \qquad \sum_{k=0}^{m} x_k, \qquad \sum_{k=0}^{m} x_k \cdot y_k, \qquad \sum_{k=0}^{m} y_k.$$

Once the normal equations are set up, then Matlab can be used to solve them. Note also that the system for the normal equation has a 2×2 coefficient matrix which is symmetric.

Example 8.1. Consider a set of data

T_k	0	1	2	3	4	5	6	7
S_k	1.15	2.32	3.32	4.53	5.65	6.97	8.02	9.23

Assume that, from physical considerations, we expect that these variables should satisfy a linear relation of the form

$$S = aT + b.$$

Use the method of least squares to find the parameters a, b that best fit the measured data.

Answer. Using (8.2.6), the normal equations are

$$\begin{cases} \displaystyle\sum_{k=0}^{m} T_k^2 \cdot a + \sum_{k=0}^{m} T_k \cdot b = \sum_{k=0}^{m} T_k S_k, \\ \displaystyle\sum_{k=0}^{m} T_k \cdot a + (m+1)\,b = \sum_{k=0}^{m} S_k. \end{cases}$$

In this case we have $m = 7$, and

$$\sum_{k=0}^{7} T_k^2 = 0^2 + 1^2 + 2^2 + \cdots + 7^2 = 140,$$

$$\sum_{k=0}^{7} T_k = \cdots = 28,$$

$$\sum_{k=0}^{7} T_k S_k = \cdots = 192.73,$$

$$\sum_{k=0}^{7} S_k = \cdots = 41.19.$$

The normal equations become

$$\begin{cases} 140\,a + 28\,b = 192.73, \\ 28\,a + 8\,b = 41.19. \end{cases}$$

Solving this linear system we obtain

$$a = 1.1563095, \qquad b = 1.1016667.$$

Therefore, the straight line that best fits the data is:

$$S(T) = 1.1563095\,T + 1.1016667.$$

In Figure 8.2 we plot the data set together with the straight line obtained by the least squares method. Observe that some data points lie above the line, and some data points lie below. For linear regression, this always happens.

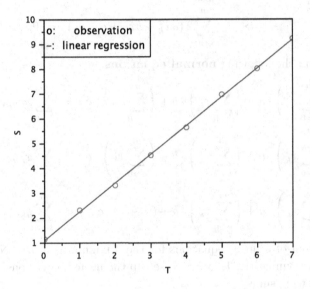

Fig. 8.2   Linear regression with least squares method.

8.3   Method of Least Squares with Quadratic Functions

Instead of a linear function, we now consider the problem of approximating a set of data with a quadratic function, i.e., with a polynomial of degree 2.

Let $m + 1$ data points $(x_k, y_k)_{k=0}^{m}$ be given. We seek three coefficients a, b, c in such a way that the quadratic polynomial $y(x) = ax^2 + bx + c$ best fits the given data.

For this purpose, we introduce the error function

$$\psi(a, b, c) \doteq \sum_{k=0}^{m} \left(ax_k^2 + bx_k + c - y_k\right)^2.$$

We wish to find values a, b, c which make this error function ψ as small as possible. At a point where the minimum is attained, the partial derivatives must vanish:

$$\frac{\partial \psi}{\partial a} = \frac{\partial \psi}{\partial b} = \frac{\partial \psi}{\partial c} = 0.$$

Writing these out, we get

$$\frac{\partial \psi}{\partial a} = 0 \quad \Longrightarrow \quad \sum_{k=0}^{m} 2 \left(ax_k^2 + bx_k + c - y_k \right) \cdot x_k^2 = 0,$$

$$\frac{\partial \psi}{\partial b} = 0 \quad \Longrightarrow \quad \sum_{k=0}^{m} 2 \left(ax_k^2 + bx_k + c - y_k \right) \cdot x_k = 0,$$

$$\frac{\partial \psi}{\partial c} = 0 \quad \Longrightarrow \quad \sum_{k=0}^{m} 2 \left(ax_k^2 + bx_k + c - y_k \right) = 0.$$

We thus obtain the following **normal equations**:

$$\begin{cases} \left(\displaystyle\sum_{k=0}^{m} x_k^4 \right) \cdot a + \left(\displaystyle\sum_{k=0}^{m} x_k^3 \right) \cdot b + \left(\displaystyle\sum_{k=0}^{m} x_k^2 \right) \cdot c = \displaystyle\sum_{k=0}^{m} x_k^2 y_k, \\[2em] \left(\displaystyle\sum_{k=0}^{m} x_k^3 \right) \cdot a + \left(\displaystyle\sum_{k=0}^{m} x_k^2 \right) \cdot b + \left(\displaystyle\sum_{k=0}^{m} x_k \right) \cdot c = \displaystyle\sum_{k=0}^{m} x_k y_k, \\[2em] \left(\displaystyle\sum_{k=0}^{m} x_k^2 \right) \cdot a + \left(\displaystyle\sum_{k=0}^{m} x_k \right) \cdot b + (m+1) \cdot c = \displaystyle\sum_{k=0}^{m} y_k. \end{cases}$$

This is a system of 3 linear equations for the 3 unknowns a, b, c. Notice again that the system is symmetric. In order to set up the normal equations, we need to compute the following sums:

$$\sum x_k^4, \quad \sum x_k^3, \quad \sum x_k^2, \quad \sum x_k, \quad \sum x_k^2 y_k, \quad \sum x_k y_k, \quad \sum y_k.$$

8.4 Method of Least Squares with General Functions (optional)

The method of least squares can also be applied in order to approximate a data set with general functions, not necessarily polynomials. We describe here a fairly general situation.

Let $m + 1$ data points $(x_k, y_k)_{k=0}^{m}$ be given, together with three functions $f(x), g(x), h(x)$. We seek three coefficients a, b, c in such a way that the function

$$y(x) = a \cdot f(x) + b \cdot g(x) + c \cdot h(x),$$

best fits the given data. Here f, g, h can be arbitrary functions, for example

$$f(x) = e^x, \qquad g(x) = \ln x, \qquad h(x) = \cos x.$$

As in the previous section, we define the error function:

$$\psi(a, b, c) \doteq \sum_{k=0}^{m} (y(x_k) - y_k)^2$$

$$= \sum_{k=0}^{m} (a \cdot f(x_k) + b \cdot g(x_k) + c \cdot h(x_k) - y_k)^2.$$

We want to find (a, b, c) minimizing the function ψ. At a point where the minimum is attained, the partial derivatives of ψ must vanish. Therefore

$$\frac{\partial \psi}{\partial a} = 0 \implies \sum_{k=0}^{m} 2 \Big[a \cdot f(x_k) + b \cdot g(x_k) + c \cdot h(x_k) - y_k \Big] \cdot f(x_k) = 0,$$

$$\frac{\partial \psi}{\partial b} = 0 \implies \sum_{k=0}^{m} 2 \Big[a \cdot f(x_k) + b \cdot g(x_k) + c \cdot h(x_k) - y_k \Big] \cdot g(x_k) = 0,$$

$$\frac{\partial \psi}{\partial c} = 0 \implies \sum_{k=0}^{m} 2 \Big[a \cdot f(x_k) + b \cdot g(x_k) + c \cdot h(x_k) - y_k \Big] \cdot h(x_k) = 0.$$

In this case, the **normal equations** are

$$\begin{cases} \left(\sum_{k=0}^{m} f(x_k)^2 \right) a + \left(\sum_{k=0}^{m} f(x_k)g(x_k) \right) b + \left(\sum_{k=0}^{m} f(x_k)h(x_k) \right) c = \sum_{k=0}^{m} f(x_k)y_k, \\[2ex] \left(\sum_{k=0}^{m} f(x_k)g(x_k) \right) a + \left(\sum_{k=0}^{m} g(x_k)^2 \right) b + \left(\sum_{k=0}^{m} h(x_k)g(x_k) \right) c = \sum_{k=0}^{m} g(x_k)y_k, \\[2ex] \left(\sum_{k=0}^{m} f(x_k)h(x_k) \right) a + \left(\sum_{k=0}^{m} g(x_k)h(x_k) \right) b + \left(\sum_{k=0}^{m} h(x_k)^2 \right) c = \sum_{k=0}^{m} h(x_k)y_k. \end{cases}$$

This is a linear system of 3 equations for the 3 unknowns a, b, c. Observe that this system is symmetric, hence we only need to compute six of the entries in the coefficient matrix.

Notice that, even if the functions $f(x), g(x), h(x)$ are not linear, the equations for a, b, c are linear. For this reason we say that this is a linear method.

8.5 General Linear Method of Least Squares

We now consider the fully general situation. Let $g_0, g_1, g_2, \cdots g_n$ be $n + 1$ be given functions (they don't need to be linear, or polynomial functions).

Given a data set (x_k, y_k), $k = 0, 1, \cdots, m$, we seek a function having the form

$$y(x) = \sum_{i=0}^{n} c_i g_i(x)$$

that best fits the data. Note that $y(x)$ depends on the parameters c_i in a linear way, therefore this will lead to a system of linear equations.

Here the functions g_i are called *basis functions*. They need to be chosen with some care.

How to choose the basis functions?   They are chosen so that the system of the normal equations is regular (invertible) and well-conditioned.

Requirements for the basis functions.   The functions $g_0, g_1, \ldots, g_n$ should be *linearly independent functions*. This means: none of the functions $g_i(x)$ can be written as linear combination of the other n functions $g_j(x)$, with $j \neq i$. An equivalent way to say that the g_i are linearly independent is:

$$\sum_{i=0}^{n} c_i g_i(x) = 0 \ \text{ for every } \ x \qquad \Longleftrightarrow \qquad c_0 = c_1 = c_2 = \cdots = c_n = 0.$$

For example, if $g_0(x) = x$ and $g_1(x) = 2x$, these two functions are not linearly independent because $g_2(x) = 2 \cdot g_1(x)$ for all x. We cannot take g_0, g_1 as basis functions.

Moreover, none of the basis functions g_i should be identically zero. (Could you think of a reason?)

We now define the error function

$$\psi(c_0, c_1, \cdots, c_n) \doteq \sum_{k=0}^{m} \Big[y(x_k) - y_k \Big]^2$$

$$= \sum_{k=0}^{m} \left[\left(\sum_{i=0}^{n} c_i g_i(x_k) \right) - y_k \right]^2.$$

We seek a point $(c_0, c_1, \ldots, c_n)$ where ψ attains its minimum. At the minimum, we have

$$\frac{\partial \psi}{\partial c_j} = 0, \qquad j = 0, 1, \cdots, n.$$

This yields $n + 1$ equations for the $n + 1$ unknowns $c_0, \ldots, c_n$:

$$\sum_{k=0}^{m} 2 \left[\left(\sum_{i=0}^{n} c_i g_i(x_k) \right) - y_k \right] g_j(x_k) = 0.$$

Rearranging terms, we obtain

$$\sum_{i=0}^{n} \left(\sum_{k=0}^{m} g_i(x_k) g_j(x_k) \right) c_i = \sum_{k=0}^{m} g_j(x_k) y_k, \qquad j = 0, 1, \cdots, n.$$

This gives the linear system of **normal equations:**

$$A\vec{c} = \vec{b},$$

where $\vec{c} = (c_0, c_1, \cdots, c_n)^t$ and

$$A = \{a_{ij}\}, \qquad a_{ij} = \sum_{k=0}^{m} g_i(x_k) g_j(x_k),$$

$$\vec{b} = \{b_j\}, \qquad b_j = \sum_{k=0}^{m} g_j(x_k) y_k.$$

We note that the coefficient matrix A is always symmetric.

8.6 Matlab Simulations for Linear Method of Leasts Squares

We go through an example where we carry out the computation in Matlab.

Example 8.2. We seek a formula describing the height $H(t)$ of the water in the North Sea, raising with the tides. We expect that this should be a periodic function of time, with period of 12 hours. It will thus have the form

$$H(t) = a_0 + a_1 \sin \frac{2\pi t}{12} + a_2 \cos \frac{2\pi t}{12},$$

for some constants a_0, a_1, a_2, to be determined. We have the following measurements of the height of the water:

t	0.0	2.0	4.0	6.0	8.0	10.0	(hours)
$H(t)$	1.0	1.6	1.4	0.6	0.2	0.8	(meters)

Which values of a_0, a_1 and a_2 yield the best fit to the data?

Answer. We set up the normal equations:

$$\sum_k \left[a_0 + a_1 \sin \frac{\pi t_k}{6} + a_2 \cos \frac{\pi t_k}{6} - H_k \right] = 0,$$

$$\sum_k \left[a_0 + a_1 \sin \frac{\pi t_k}{6} + a_2 \cos \frac{2\pi t_k}{12} - H_k \right] \sin \frac{\pi t_k}{6} = 0,$$

$$\sum_k \left[a_0 + a_1 \sin \frac{\pi t_k}{6} + a_2 \cos \frac{2\pi t_k}{12} - H_k \right] \cos \frac{\pi t_k}{6} = 0,$$

i.e.,

$$a_0(n+1) + a_1 \sum_k \sin \frac{\pi t_k}{6} + a_2 \sum_k \cos \frac{\pi t_k}{6} = \sum_k H_k,$$

$$a_0 \sum_k \sin \frac{\pi t_k}{6} + a_1 \sum_k \sin^2 \frac{\pi t_k}{6} + a_2 \sum_k \cos \frac{\pi t_k}{6} \sin \frac{\pi t_k}{6}$$

$$= \sum_k H_k \sin \frac{\pi t_k}{6},$$

$$a_0 \sum_k \cos \frac{\pi t_k}{6} + a_1 \sum_k \sin \frac{\pi t_k}{6} \cos \frac{\pi t_k}{6} + a_2 \sum_k \cos^2 \frac{\pi t_k}{6}$$

$$= \sum_k H_k \cos \frac{\pi t_k}{6}.$$

Simple Matlab codes: To work out this problem using Matlab, we need to (i) compute the various sums, (ii) set up the linear system of normal equations, and (iii) solve the system and plot the result. Here is a possible list of commands:

```
t=[0 2 4 6 8 10];
H=[1 1.6 1.4 0.6 0.2 0.8];

n=length(t);
va=pi/6;

s1=sum(sin(va*t));
s2=sum(cos(va*t));
s3=sum(sin(va*t).^2);
s4=sum(cos(va*t).*sin(va*t));
s5=sum(cos(va*t).^2);

A=[n,s1,s2; s1, s3, s4; s2, s4, s5];
h=[sum(H);sum(H.*sin(va*t));sum(H.*cos(va*t))];

a=A\h;

x=[0:0.05:12];
fx=a(1)+a(2)*sin(va*x)+a(3)*cos(va*x);

plot(x,fx,'b',t,H,'ro')
```

The code gives:

$$a_0 = 0.9333, \quad a_1 = 0.5774, \quad a_2 = 0.2667.$$

The plot is shown in Figure 8.3.

8.7   Non-linear Methods of Least Squares

To illustrate these methods, we present two non-linear examples. The first one is called quasi-linear, which becomes linear after a variable change. The second one is truly non-linear.

Example 8.3. Let the data points (t_k, y_k) be given, for $k = 0, 1, 2, \cdots, m$. Assume that, by some modeling considerations, we believe these data should follow a law of the form

$$y(t) \; = \; a \cdot b^t.$$

For example, this could be a model of growth, where $y(t)$ is the size of a population at time t, a is the initial size, and b describes a growth rate. We wish to find coefficients a, b so that the function $y(t)$ best fits the given data.

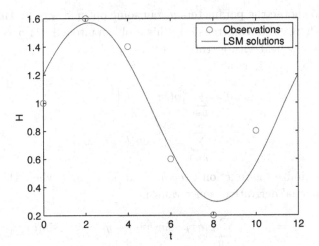

Fig. 8.3   Least squares method for ocean waves.

We first make a variable change:

$$\ln y = \ln a + t \cdot \ln b,$$

and let

$$S = \ln y, \qquad \bar{a} = \ln a, \qquad \bar{b} = \ln b,$$

which leads to the function

$$S(x) = \bar{a} + \bar{b}t.$$

Since S depends on $(\bar{a}, \bar{b})$ in a linear way, this will lead to a linear least squares method.

As a first step, we generate a new data set (t_k, S_k) where $S_k = \ln y_k$ for all k.

Then, we use linear least squares method and find the parameters $(\bar{a}, \bar{b})$ such that $S(x)$ best fits (t_k, S_k).

Finally, we go back to the original variables:

$$a = \exp\{\bar{a}\}, \qquad b = \exp\{\bar{b}\}.$$

Example 8.4. A truly non-linear problem.   Let the data points (x_k, y_k) be given, for $k = 0, 1, 2, \cdots, m$. We seek a function of the form

$$y(x) = (ax) \cdot \sin(bx),$$

which best fits the given data. Functions of this form arise in the solution of a second order ODE with resonance. Here a is the rate at which the amplitude grows, and b is the frequency.

We want to choose the parameters a, b in some optimal way. In this case, there is no variable change that can transform this problem into a linear one (with respect to a, b). So we now deal with a truly non-linear problem.

We define the error function

$$\psi(a, b) = \sum_{k=0}^{m} \left[y(x_k) - y_k \right]^2$$

$$= \sum_{k=0}^{m} \left[a x_k \cdot \sin(b x_k) - y_k \right]^2.$$

We want to minimize the function $\psi(a, b)$. At a point (a, b) where the minimum is attained, the partial derivatives must vanish:

$$\frac{\partial \psi}{\partial a} = 0 : \quad \Longrightarrow \quad \sum_{k=0}^{m} 2 \left[a x_k \cdot \sin(b x_k) - y_k \right] \cdot \left[x_k \cdot \sin(b x_k) \right] = 0 \,,$$

$$\frac{\partial \psi}{\partial b} = 0 : \quad \Longrightarrow \quad \sum_{k=0}^{m} 2 \left[a x_k \cdot \sin(b x_k) - y_k \right] \cdot \left[a x_k \cdot \cos(b x_k) x_k \right] = 0 \,.$$

We now have a system of two non-linear equations to solve for the two variables a, b.

We may use Newton's method for nonlinear systems to find a root.

Be aware that this system may have several solutions, including all the points where ψ attains a minimum, or a maximum, or has a saddle. Once you find an approximate solution to the normal equations, you need to run the 2-nd derivative test to check if it is really a minimum. As you can imagine, this is a much harder problem than the linear one!

8.8   Method of Least Squares for Continuous Functions

In the previous sections, given a set of points (x_k, y_k), we approximated these data with a function $y(x)$ having a prescribed form.

In this section, given a function $y = f(x)$, we shall approximate it with a linear combination of other functions g_i having a prescribed form.

Problem setting.   Let a function $f(x)$ be given, defined on the interval $x \in [a, b]$, together with n other functions $g_1(x), \ldots, g_n(x)$. We seek a function $g(x)$ of the form

$$g(x) = \sum_{i=1}^{n} a_i g_i(x) \,,$$

such that the error

$$E(f, g) \doteq \| f - g \|_2^2 = \int_a^b (f(x) - g(x))^2 \, dx$$

is minimized.

Here the functions g_i, are called the *basis functions*. For example, we may choose

$$\{g_0(x),\, g_1(x),\, g_2(x),\, \ldots,\, g_n(x)\} \;=\; \{1, x, x^2, \ldots, x^n\}.$$

In this case, our problem is to approximate $f(x)$ with a polynomial of degree n, in the best possible way.

Other choices of the functions g_i are also possible. The main requirement is that the set of basis functions $\{g_i\}$ should be linearly independent.

Notice that the error $E(f,g)$ is a function of the coefficients a_i. We can write

$$E(f,g) \;=\; E(a_1, a_2, \cdots, a_n) \;=\; \int_a^b \left(f(x) - \sum_{i=1}^n a_i g_i(x) \right)^2 dx.$$

We want to find the coefficients $(a_1, a_2, \cdots, a_n)$ in order to minimize this error. At a point where the minimum is attained, all partial derivatives must vanish:

$$\frac{\partial E}{\partial a_i} = 0, \qquad i = 1, 2, \ldots, n. \tag{8.8.7}$$

Since

$$\frac{\partial E}{\partial a_i} = -2 \int_a^b g_i(x) \left(f(x) - \sum_{j=1}^n a_j g_j(x) \right) dx$$

$$= -2 \int_a^b g_i(x) f(x)\, dx + 2 \int_a^b g_i(x) \sum_{j=1}^n a_j g_j(x)\, dx,$$

the condition (8.8.7) becomes

$$\sum_{j=1}^n a_j \int_a^b g_i(x) g_j(x)\, dx = \int_a^b g_i(x) f(x)\, dx, \qquad i = 1, 2, \ldots, n.$$

Note that we have a system of n linear equations for the n unknowns $a_1, \ldots, a_n$. Writing these equations into matrix-vector form, we obtain

$$C\vec{a} = \vec{b},$$

where C is the coefficient matrix

$$C = (c_{ij}), \qquad c_{ij} = \int_a^b g_i(x) g_j(x)\, dx,$$

and $\vec{b}$ is the load vector

$$\vec{b} = (b_i), \qquad b_i = \int_a^b g_i(x) f(x)\, dx.$$

Note that the matrix C is symmetric, because

$$c_{ij} = \int_a^b g_i(x) g_j(x)\, dx = \int_a^b g_j(x) g_i(x)\, dx = c_{ji}.$$

Well posedness. If the basis functions $\{g_i\}$ are linearly independent, then the matrix C is non-singular. Hence the problem is well-posed, and the system $C\vec{a} = \vec{b}$ has a unique solution.

Some particular choices of $\{g_i\}$ lead to very interesting and elegant constructions.

Orthogonal basis. If the basis functions $g_1(x), \ldots, g_n(x)$ are orthogonal to each other, i.e., if

$$\int_a^b g_i(x)g_j(x)\,dx = 0 \quad \text{whenever } i \neq j,$$

then the matrix C is diagonal. Indeed

$$c_{ii} = \int_a^b (g_i(x))^2\,dx, \qquad c_{ij} = 0 \quad \text{for } i \neq j.$$

In this special case, the system is easy to solve. We have

$$a_i = \frac{b_i}{a_{ii}} = \frac{\int_a^b g_i(x)f(x)\,dx}{\int_a^b (g_i(x))^2\,dx}.$$

We now review some famous examples of families of orthogonal functions.

Legendre polynomials: For the interval $x \in [-1, 1]$, these polynomials are:

$$P_0(x) = 1,$$
$$P_1(x) = x,$$
$$P_2(x) = (3x^2 - 1)/2,$$
$$P_3(x) = (5x^3 - 3x)/2,$$
$$P_4(x) = (35x^4 - 30x^2 + 3)/8,$$

$$\cdots.$$

They can be obtained as solutions to the Legendre equation, which we shall study in a later chapter.

We now explore some properties of these polynomials.

Example 8.5. Verify that $P_0 = 1, P_1 = x, P_2 = (3x^2 - 1)/2$ are orthogonal to each other on the interval $[-1, 1]$.

Answer. Note that P_0, P_2 are even functions, and P_1 is an odd function.

Then, the products $P_0(x)P_1(x)$ and $P_1(x)P_2(x)$ are odd functions. So we immediately have

$$\int_{-1}^1 P_0(x)P_1(x)\,dx = 0,$$

$$\int_{-1}^1 P_1(x)P_2(x)\,dx = 0,$$

indicating that P_0, P_1 are orthogonal to each other, and so are P_1 and P_2. It remains to check that P_0 and P_2 are orthogonal. We compute

$$\int_{-1}^{1} P_0(x)P_2(x)\,dx = \frac{1}{2}\int_{-1}^{1}(3x^2-1)\,dx$$
$$= \frac{1}{2}(x^3-x)\Big|_{-1}^{1}$$
$$= 0.$$

Hence the three polynomials P_0, P_1, P_2 are orthogonal to each other.

Example 8.6. Let P_0, P_1, P_2 be the Legendre polynomials in the previous example. Find a function

$$g(x) \doteq a_0 P_0(x) + a_1 P_1(x) + a_2 P_2(x), \qquad \text{on} \quad -1 \le x \le 1,$$

that "best" approximates the following function in the least squares sense

$$f(x) = \begin{cases} -1, & -1 \le x \le 0, \\ 1, & 0 < x \le 1. \end{cases}$$

Answer. Our goal is to find the coefficients (a_0, a_1, a_2). Since $\{P_0, P_1, P_2\}$ is an orthogonal basis, the coefficients a_i are simply computed as

$$a_i = \frac{\int_{-1}^{1} f(x)P_i(x)\,dx}{\int_{-1}^{1} P_i^2(x)\,dx}.$$

Note that, since $f(x)$ is odd, the product functions $f(x)P_0(x)$ and $f(x)P_2(x)$ are odd. Therefore, their integrals over $[-1, 1]$ are zero. This gives $a_0 = 0$, $a_2 = 0$.

It remains to compute a_1. We have

$$a_1 = \frac{\int_{-1}^{1} f(x)P_1(x)\,dx}{\int_{-1}^{1} P_1^2(x)\,dx}$$
$$= \frac{2\int_{0}^{1} x\,dx}{\int_{-1}^{1} x^2\,dx}$$
$$= \frac{2(0.5)}{2/3} = \frac{3}{2}.$$

Therefore, the "best" approximation is

$$g(x) = \frac{3}{2}P_1(x) = \frac{3}{2}x.$$

As a further example, we recall that the trigonometric functions

$$1, \quad \sin nx, \quad \cos nx, \quad n = 1, 2, \cdots,$$

are orthogonal to each other, on the interval $x \in [-\pi, \pi]$. This means,

$$\int_{-\pi}^{\pi} 1 \cdot \sin nx \, dx = 0, \qquad \text{for all } n \geq 1,$$

$$\int_{-\pi}^{\pi} 1 \cdot \cos nx \, dx = 0, \qquad \text{for all } n \geq 1,$$

$$\int_{-\pi}^{\pi} \sin nx \cdot \sin mx \, dx = 0, \qquad \text{for all } m \neq n,$$

$$\int_{-\pi}^{\pi} \cos nx \cdot \cos mx \, dx = 0, \qquad \text{for all } m \neq n,$$

$$\int_{-\pi}^{\pi} \sin nx \cdot \cos mx \, dx = 0, \qquad \text{for all } m, n.$$

You are encouraged to carry out the integrations and verify these properties!

Example 8.7. Let $f(x)$ be a function defined on $[-\pi, \pi]$. Given an integer M, we want to approximate the function f by a sum of trigonometric functions

$$g(x) = c_0 + \sum_{n=1}^{M} \left[a_n \sin nx + b_n \cos nx \right],$$

in the "best" possible way.

Answer. Using the orthogonality property, the coefficients are computed as

$$a_n = \frac{\int_{-\pi}^{\pi} f(x) \sin nx \, dx}{\int_{-\pi}^{\pi} \sin^2 nx \, dx} = \frac{1}{\pi} \int_{-\pi}^{\pi} f(x) \sin nx \, dx,$$

$$b_n = \frac{\int_{-\pi}^{\pi} f(x) \cos nx \, dx}{\int_{-\pi}^{\pi} \cos^2 nx \, dx} = \frac{1}{\pi} \int_{-\pi}^{\pi} f(x) \cos nx \, dx,$$

$$c_0 = \frac{\int_{-\pi}^{\pi} f(x) \cdot 1 \, dx}{\int_{-1}^{1} 1 \, dx} = \frac{1}{2\pi} \int_{-\pi}^{\pi} f(x) \, dx.$$

Notice that these are precisely the coefficients of the *Fourier series* for the function $f(x)$.

8.9 Homework Problems for Chapter 8

1. Simplest Problem Using the Least Squares Method

Use the method of least squares, find the constant function that best fits the data

x	-1	2	3
y	5/4	4/3	5/12

2. The Method of Least Squares with Polynomial Regression

(a). What straight line best fits the following data in the least-squares sense?

x	1	2	3	4
y	0	1	1	2

(b). Find the equation of a parabola of form $y = ax^2 + b$ that best represents the following data, in least-squares sense:

x	-1	0	1
y	3.1	0.9	2.9

3. The Method of Least Squares with Non-polynomial Functions

(a). We are given a data set (x_k, y_k), with $k = 0, \ldots, m$. We seek a function of the form

$$g(x) = \alpha \sin x + \beta \cos x$$

that best approximates the data. Set up the normal equations, which solve the problem with the method of least squares.

Compute the values of α and β which provide the best fit to the particular data

x	1.0	1.5	2.0	2.5
y	1.902	0.5447	-0.9453	-2.204

(b). Let $f(x)$ be a given function. What constant c makes the expression

$$\sum_{k=0}^{m} [f(x_k) - ce^{x_k}]^2$$

as small as possible?

4. Application: A Population Model

Assume that the growth of the world's population could be described by the ODE

$$\frac{d}{dt} p(t) = K p^2(t),$$

where

$$p = \text{world population, in millions of people,}$$
$$t = \text{time, measured in years,}$$
$$K = \text{growth rate.}$$

The general solution of this ODE has the form

$$p(t) = \frac{1}{K} \cdot \frac{1}{t_0 - t} \qquad (*)$$

for some constants K, t_0. Throughout the past, we have recorded the following historical data:

t	1650	1700	1750	1800	1850	1900	1920	1940	1960
p	545	623	728	906	1171	1608	1834	2295	3003

(a). Consider the new variables $y = 10^6/p$, and $x = t - 1830$. Rewrite the equation
     (*) in the form $y(x) = a_0 + a_1 x$.
(b). Use the method of least squares to determine a_0 and a_1.
(c). Going back to the original variables, find K and t_0. Now you can write out the
     function $p(t)$ that describes the population p as a function of time t.
(d). Use the model to predict the population for $t = 1980$, $t = 2000$ and $t = 2010$.
     Are you happy with the result? Comparing to the actual population in 2011,
     which is about 7.1 billions, what would you say about this model?

5. Quasi-linear Method of Least Squares

For the table

x_k	0.0	0.2	0.4	0.6	0.8	1.0
y_k	1.996	1.244	0.810	0.541	0.375	0.259

we seek a least squares approximation of the form

$$\bar{y}(x) = a_0 \cdot \frac{e^{a_1 x}}{(1+x)^{a_2}},$$

where a_0, a_1 and a_2 are three unknown coefficients.

(a). Write the normal equations for your approximation.
     Try to reformulate the problem in a suitable way, so that, after a variable
     change, you get a linear system.
(b). Use Matlab to compute the solution to the normal equations. Make a plot which
     shows the data points and the least squares approximation $\bar{y}(x)$. Compute the
     sum of squares

$$\sum_k (y_k - \bar{y}(x_k))^2.$$

6. The Method of Least Squares in Matlab

Given data set

x_k	0	0.1	0.2	0.3	0.4	0.5	0.6	0.7	0.8	0.9	1
y_k	0.7829	0.8052	0.5753	0.5201	0.3783	0.2923	0.1695	0.0842	0.0415	0.009	0

Use Matlab function `polyfit` to find respectively the first, second, fourth and eighth order polynomials that best fit the data, using the method of least squares. You should also plot your polynomials, together with the data set.

What to hand in: The script file, the plots, and your comments.

7. Least Squares Approximation of Functions

Let $f(x)$ be a function defined on the interval $[-1, 1]$, as

$$f(x) = \begin{cases} -1, & -1 \le x < 0, \\ 1, & 0 \le x \le 1. \end{cases}$$

We want to approximate $f(x)$ by a function of the form

$$g(x) = a\cos(\pi x) + b\sin(\pi x).$$

Find the best possible constants a and b, by the method of least squares.

Chapter 9

Numerical Solution of ODEs

9.1 Introduction

In this chapter, we study numerical methods for ODEs (Ordinary Differential Equations), starting with the initial value problem. Roughly speaking, an ODE is an equation which contains one or more derivatives of the unknown function (of one single variable).

If the equation involves only a first derivative, we say that the ODE is of first order. If it contains the first and second derivatives, we say that the ODE is of second order.

Example 9.1. Let $x = x(t)$ be the unknown function of t. Examples of ODEs are:

$$x' = 3x + 2, \qquad x' = x^2, \qquad x'' + x \cdot x' + 4 = 0, \qquad \text{etc.}$$

We shall focus on the IVP *(Initial Value Problem)* for a first-order ODE.

$$\begin{cases} x' = f(t, x) & \text{(differential equation),} \\ x(t_0) = x_0 & \text{(initial condition).} \end{cases} \tag{9.1.1}$$

Here are some examples of initial value problems, where we also provide the solutions.

$$x'(t) = x + 1, \quad x(0) = 0. \qquad \text{Solution:} \quad x(t) = e^t - 1.$$
$$x'(t) = 2, \quad x(1) = 0. \qquad \text{Solution:} \quad x(t) = 2t - 2.$$
$$x'(t) = 2t, \quad x(0) = 3. \qquad \text{Solution:} \quad x(t) = t^2 + 3.$$

One can easily verify that the given functions are indeed solutions, by checking that the differential equation as well as the initial condition are satisfied.

In a course on differential equations at sophomore level, various techniques are introduced to find the exact solutions for certain types of ODE. However, an explicit formula for the exact solution can be found only in special cases: linear equations (mainly with constant coefficients), separable equations, and equations that can be reduced to these simpler forms after suitable changes of variables. In the general case, exact solutions can be very difficult or even impossible to obtain. The focus of this chapter is how to construct approximate solutions, by numerical methods.

Numerical solutions: We wish to compute an approximate value for the solution at discrete sampling points. These points will be called *grid points*. The collection of these points is called *the grid*. There are many way of making a grid. The simplest one is the *uniform grid*, which we define now.

Uniform grid for time variable. Let $h > 0$ be the common length of every time step. We then define the times
$$t_k = t_0 + kh, \qquad t_0 < t_1 < \cdots < t_N.$$
Here t_0 is the initial time, t_N is the final time, $h = t_{k+1} - t_k$ is also called the *grid size*, and N is the total number of steps.

Given the ODE (9.1.1), and a time interval $[t_0, t_N]$, we wish to compute the values
$$x_n \approx x(t_n), \qquad \text{for} \quad n = 1, 2, \cdots, N.$$
These will provide approximate values of the unknown function $x(t)$ at the times $t_0 < t_1 < \cdots < t_N$.

Here is an overview of this chapter, with a list of the methods we shall cover:

- Taylor series method, and error estimates,
- Runge-Kutta methods,
- Multi-step methods,
- System of ODE,
- High order equations and systems,
- Stiff systems,
- Matlab solvers.

9.2 Taylor Series Methods for ODEs

The Taylor series methods are based on Taylor approximations, as the name suggests. The order of the method depends on how many terms in the Taylor series we use in the approximation.

Consider the IVP
$$x'(t) = f(t, x(t)), \qquad x(t_0) = x_0.$$
Let $t_1 = t_0 + h$. Let us find the value $x(t_1) = x(t_0 + h)$. The Taylor series for the function $x(t)$, expanded at $t = t_0$, can be written as
$$x(t_0 + h) = x(t_0) + hx'(t_0) + \frac{1}{2}h^2 x''(t_0) + \cdots$$
$$= \sum_{m=0}^{\infty} \frac{1}{m!} h^m x^{(m)}(t_0),$$

where $x^{(m)}$ denotes the m-th derivative.

Taylor series method of order m: We take the first $(m + 1)$ terms in Taylor expansion, i.e., we take the partial sum of the series up to the term containing $x^{(m)}$. This will be used as the approximation of x_1, as

$$x_1 \approx x(t_0 + h)$$
$$\approx x(t_0) + hx'(t_0) + \frac{1}{2}h^2x''(t_0) + \cdots + \frac{1}{m!}h^mx^{(m)}(t_0).$$

By the Taylor Theorem (in Chapter 1), the error at this step is

$$e_1 = x(t_0 + h) - x_1$$
$$= \sum_{k=m+1}^{\infty} \frac{1}{k!}h^kx^{(k)}(t_0)$$
$$= \frac{1}{(m+1)!}h^{m+1}x^{(m+1)}(\xi),$$

for some $\xi \in (t_0, t_1)$.

Having determined the value $x_1 = x(t_1)$, in order to compute the next value $x_2 = x(t_1 + h)$ we repeat the same procedure, computing the Taylor approximation to $x(t)$ at the point $t = t_1$, etc.

We now show in detail how to implement this method for $m = 1$ and $m = 2$.

Taylor series method of order 1. For $m = 1$, we have the **forward Euler step**:

$$x_1 = x_0 + hx'(t_0) = x_0 + h \cdot f(t_0, x_0).$$

The general formula for the k-th step is:

$$x_{k+1} = x_k + h \cdot f(t_k, x_k), \qquad k = 0, 1, 2, \cdots, N - 1.$$

Taylor series method of order 2. For $m = 2$, we have

$$x_1 = x_0 + hx'(t_0) + \frac{1}{2}h^2x''(t_0).$$

Using the identities

$$x''(t_0) = \frac{d}{dt}x'(t_0)$$
$$= \frac{d}{dt}f(t_0, x(t_0))$$
$$= f_t(t_0, x_0) + f_x(t_0, x_0) \cdot x'(t_0)$$
$$= f_t(t_0, x_0) + f_x(t_0, x_0) \cdot f(t_0, x_0),$$

we obtain

$$x_1 = x_0 + hf(t_0, x_0) + \frac{1}{2}h^2\left[f_t(t_0, x_0) + f_x(t_0, x_0) \cdot f(t_0, x_0)\right].$$

For the general k-th step, we have

$$x_{k+1} = x_k + hf(t_k, x_k) + \frac{1}{2}h^2 \left[f_t(t_k, x_k) + f_x(t_k, x_k) \cdot f(t_k, x_k) \right].$$

It should now be clear how one can derive a Taylor series method of order $m \geq 3$.

Example 9.2. Set up Taylor series methods with $m = 1$ and with $m = 2$ for the IVP

$$x' = -x + e^{-t}, \qquad x(0) = 0.$$

Let $h = 0.1$, and compute 2 steps to find the approximations $x_1 \approx x(0.1), x_2 \approx x(0.2)$. Comparing the results with the exact solution $x(t) = te^{-t}$, compute the errors in the approximations.

Answer. The initial data is $t_0 = 0$, $x_0 = 0$.

For $m = 1$, the first order Taylor iteration scheme is

$$x_{k+1} = x_k + h\left(-x_k + e^{-t_k}\right) = (1 - h)x_k + he^{-t_k}.$$

We take 2 steps with $h = 0.1$, so that $t_1 = 0.1$, $t_2 = 0.2$. This yields

$$x_1 = x_0 + (0.1)\left(-x_0 + e^{-t_0}\right) = 0.1,$$
$$x_2 = x_1 + (0.1)\left(-x_1 + e^{-t_1}\right) = 0.1804837.$$

Comparing these results with the exact solution, we find that the errors are

$$e_1 = |x_1 - t_1 e^{t_1}| = 0.0095163, \qquad e_2 = |x_2 - t_2 e^{t_2}| = 0.0167376.$$

For $m = 2$, we first compute

$$
\begin{aligned}
x'' &= (-x + e^{-t})' \\
&= -x' - e^{-t} \\
&= -(-x + e^{-t}) - e^{-t} \\
&= x - 2e^{-t}.
\end{aligned}
$$

Therefore the second order Taylor iteration scheme is

$$
\begin{aligned}
x_{k+1} &= x_k + hx'_k + \frac{1}{2}h^2 x''_k \\
&= x_k + h\left(-x_k + e^{-t_k}\right) + \frac{1}{2}h^2\left(x_k - 2e^{-t_k}\right) \\
&= \left(1 - h + \frac{1}{2}h^2\right)x_k + \left(h - h^2\right)e^{-t_k}.
\end{aligned}
$$

We take 2 steps with $h = 0.1$, and obtain

$$x_1 = \left(1 - h + \frac{1}{2}h^2\right)x_0 + \left(h - h^2\right)e^{-t_0} = 0.09,$$
$$x_2 = \left(1 - h + \frac{1}{2}h^2\right)x_1 + \left(h - h^2\right)e^{-t_1} = 0.1628854.$$

In this case, the errors are

$$e_1 = 0.0004837, \qquad e_2 = 0.0008608.$$

We see that the errors obtained by the second order method are much smaller than those for the first order method.

The simulation result is plotted in Figure 9.1, where we plot the exact solution together with the two approximate solutions. We see that $m = 2$ gives much better approximation than $m = 1$.

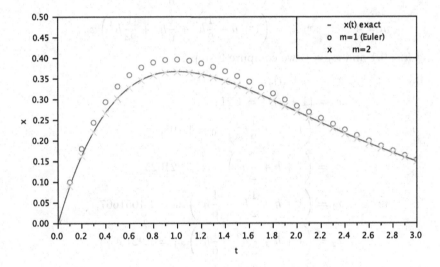

Fig. 9.1   Taylor series method with $m = 1$ and $m = 2$.

Example 9.3. Set up the Taylor series methods with $m = 1, 2, 3, 4$ for the IVP

$$x' = x, \qquad x(0) = 1.$$

Note that the exact solution is $x(t) = e^t$.

Use $h = 0.1$ and compute 2 steps. Compute also the errors.

Answer. We set $x_0 = 1$. Differentiating the equation $x' = x$ several times, we obtain

$$x'' = x' = x,$$
$$x''' = x'' = x,$$
$$\cdots$$
$$x^{(m)} = x.$$

The Taylor series method of order m thus takes the form:

$$x_{k+1} = x_k + hx_k + \frac{1}{2}h^2 x_k + \cdots + \frac{1}{m!}h^m x_k$$

$$= \left(1 + h + \frac{1}{2}h^2 + \frac{1}{6}h^3 + \cdots + \frac{1}{m!}h^m\right) x_k.$$

Now it is easy to write out the Taylor series methods for various values of m:

$$m = 1: \qquad x_{k+1} = (1+h)x_k,$$

$$m = 2: \qquad x_{k+1} = \left(1 + h + \frac{1}{2}h^2\right) x_k,$$

$$m = 3: \qquad x_{k+1} = \left(1 + h + \frac{1}{2}h^2 + \frac{1}{6}h^3\right) x_k,$$

$$m = 4: \qquad x_{k+1} = \left(1 + h + \frac{1}{2}h^2 + \frac{1}{6}h^3 + \frac{1}{24}h^4\right) x_k.$$

Using $h = 0.1$ and $x_0 = 1$, we compute 2 steps:

$$m = 1: \qquad x_1 = (1+h)x_0 = 1.1,$$

$$x_2 = (1+h)x_1 = 1.21,$$

$$m = 2: \qquad x_1 = \left(1 + h + \frac{1}{2}h^2\right) x_0 = 1.105,$$

$$x_2 = \left(1 + h + \frac{1}{2}h^2\right) x_1 = 1.221025,$$

$$m = 3: \qquad x_1 = \left(1 + h + \frac{1}{2}h^2 + \frac{1}{6}h^3\right) x_0 = 1.1051667,$$

$$x_2 = \left(1 + h + \frac{1}{2}h^2 + \frac{1}{6}h^3\right) x_1 = 1.2213934,$$

$$m = 4: \qquad x_1 = \left(1 + h + \frac{1}{2}h^2 + \frac{1}{6}h^3 + \frac{1}{24}h^4\right) x_0 = 1.1051708,$$

$$x_2 = \left(1 + h + \frac{1}{2}h^2 + \frac{1}{6}h^3 + \frac{1}{24}h^4\right) x_1 = 1.2214026.$$

The errors are

$$m = 1: \qquad e_1 = 5.2 \cdot 10^{-3}, \qquad e_2 = 1.1 \cdot 10^{-2},$$

$$m = 2: \qquad e_1 = 1.7 \cdot 10^{-4}, \qquad e_2 = 3.8 \cdot 10^{-4},$$

$$m = 3: \qquad e_1 = 4.3 \cdot 10^{-6}, \qquad e_2 = 9.4 \cdot 10^{-6},$$

$$m = 4: \qquad e_1 = 8.5 \cdot 10^{-8}, \qquad e_2 = 2.0 \cdot 10^{-7}.$$

We see a clear decrease in the errors, when higher order methods are used.

The results of the simulation are included in Figure 9.2, where we plot the exact solution together with the 4 approximate solutions. We see that $m \geq 2$ gives a much better approximation than $m = 1$.

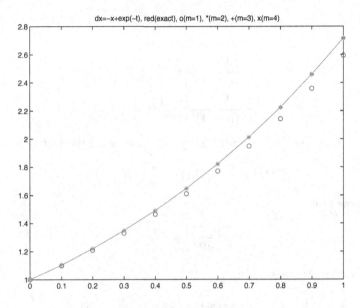

Fig. 9.2   Taylor series method with $m = 1, 2, 3, 4$.

Error analysis.   We consider the IVP for the ODE

$$x' = f(t, x), \qquad x(t_0) = x_0. \tag{9.2.2}$$

The *local truncation error* is the error made at each time step, by truncating the Taylor series taking only finitely many terms. A precise definition is given below.

Definition 9.1. Let t_k, x_k be given, let x_{k+1} be the numerical solution after one iteration, and let $x(t_k + h)$ be the exact solution for the IVP

$$x' = f(t, x), \qquad x(t_k) = x_k.$$

Then the *local truncation error* is defined as

$$e_k \doteq |x_{k+1} - x(t_k + h)|.$$

The following estimate on the local truncation error is a building block for all further error analysis.

Theorem 9.1. *For the Taylor series method of order m, at each step k the local error is of order $m+1$, i.e., $e_k \leq Mh^{m+1}$ for some constant M which is independent of h.*

Sketch of the proof.   Since $x(t)$ is a function of t, the composite function $f(t, x(t))$ can be treated as a function of a single variable t. By Taylor's theorem we have

$$\begin{aligned}
e_k &= |x_{k+1} - x(t_k + h)| \\
&= \frac{h^{m+1}}{(m+1)!} \left| x^{(m+1)}(\xi) \right| \\
&= \frac{h^{m+1}}{(m+1)!} \left| \frac{d^m f}{dt^m}(\xi, x(\xi)) \right|,
\end{aligned}$$

for some $\xi \in (t_k, t_{k+1})$. In the above computation we used the fact that

$$x^{(m+1)} = \frac{d^m}{dt^m}(x'(t)) = \frac{d^m}{dt^m} f(t, x(t)).$$

Assuming now that

$$\left| \frac{d^m f}{dt^m} \right| \le \widehat{M},$$

for some constant $\widehat{M}$, we obtain

$$e_k \le \frac{\widehat{M}}{(m+1)!} h^{m+1} = M h^{m+1} = \mathcal{O}(h^{m+1}), \qquad M = \frac{\widehat{M}}{(m+1)!}.$$

$\square$

Total error. The total error is the error accumulated at the terminal time T. More precisely, assume we want to compute $x(T)$ at some time $t = T$. We divide the interval $[0, T]$ into N equal parts, choosing a time step $h = T/N$. Then the total number of steps is

$$N = \frac{T}{h}, \qquad \text{hence} \quad T = Nh.$$

The total error is defined as

$$E \doteq |x(T) - x_N|.$$

In order to derive an estimate on the total error, we must assume that the ODE is well-posed.

Definition 9.2. We say that the system (9.2.2) is *well-posed* on the interval $[t_0, T]$ if, for some constant C, the following holds. Let $x(t)$ be the solution for (9.2.2), and let $\tilde{x}(t)$ be the solution for (9.2.2) with a different initial data $\tilde{x}(t_0) = \tilde{x}_0$. Then

$$|x(t) - \tilde{x}(t)| \le C |x_0 - \tilde{x}_0| \qquad \text{for all } t \in [t_0, T].$$

In other words, if we make a small perturbation of the initial data, the solution will not change much. For such well-posed ODE, we have the following error estimate.

Theorem 9.2. *Assume that the system (9.2.2) is well-posed. If the local error of a numerical iteration satisfies*

$$e_k \leq Mh^{m+1},$$

then the total error satisfies

$$E \leq Ch^m,$$

for some bounded constant C, where C does not depend on h.

This means that if the local error is of order $m + 1$, then the total error is of order m.

Sketch of the proof. We observe two facts about the errors. First, at every step k, the local error is being carried on through the rest of the simulation. Second, the local errors accumulate through time iteration steps. Therefore we need to estimate how errors are generated, and how they grow in time.

Thanks to the well-posedness assumption, at each time step k, the local error e_k is amplified at most by a factor of C in the answer at the final time T. See Figure 9.3 for an illustration for e_1, e_2. The other local errors are all estimated in the same way.

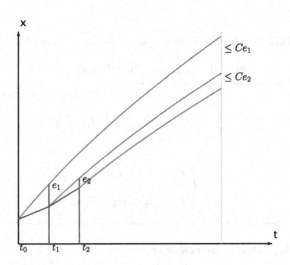

Fig. 9.3   Error evolution and accumulation in a numerical method for ODEs.

Then, we can add up all the accumulated errors at T caused by all the local errors

$$E = C \sum_{k=1}^{N} \left| e_L^{(k)} \right| \leq C \sum_{k=1}^{N} \frac{M}{(m+1)!} h^{m+1}$$

$$= CN \frac{M}{(m+1)!} h^{m+1}$$

$$= C(Nh) \frac{M}{(m+1)!} h^m$$

$$= \frac{CMT}{(m+1)!} h^m = \mathcal{O}(h^m).$$

Therefore, the method is of order m. $\qquad\qquad\qquad\qquad\qquad\square$

9.3   Runge-Kutta Methods

High order Taylor series methods require the computation of high order derivatives $x'', x''', \ldots$ of the unknown function. To obtain these derivatives, one needs to differentiate the equation several times, producing rather complicated formulas. This is a major drawback.

A better method should only use the values of the function $f(t,x)$, not its derivatives.

This leads to the so-called Runge-Kutta methods. We refer to them as RK methods.

1st order RK method.   It is the same as the forward Euler's method.

2nd order RK method.   This is called *Heun's method*.

Let $h = t_{k+1} - t_k$. Given x_k, the next value x_{k+1} is computed as

$$x_{k+1} = x_k + \frac{1}{2}(K_1 + K_2),$$

where

$$\begin{cases} K_1 = h \cdot f(t_k, x_k), \\ K_2 = h \cdot f(t_k + h, x_k + K_1). \end{cases}$$

Theorem 9.3. *Heun's method is of second order.*

Proof. Thanks to Theorem 9.2, it suffices to show that the local truncation error is of order 3.

The Taylor expansion of f (which is a function of two variables) gives

$$f(t_k + h, x_k + K_1) = f(t_k, x_k) + h f_t(t_k, x_k) + K_1 f_x(t_k, x_k) + \mathcal{O}(h^2, K_1^2).$$

We have $K_1 = hf(t_k, x_k)$, so the last term above is actually $\mathcal{O}(h^2)$. We also have

$$K_2 = h\Big[f(t_k, x_k) + hf_t(t_k, x_k) + hf(t_k, x_k)f_x(t_k, x_k) + \mathcal{O}(h^2)\Big].$$

Then, our method is: (we drop the dependence of (t_k, x_k) below)

$$\begin{aligned}
x_{k+1} &= x_k + \frac{1}{2}\Big[hf + hf + h^2 f_t + h^2 ff_x + \mathcal{O}(h^3)\Big] \\
&= x_k + hf + \frac{1}{2}h^2[f_t + ff_x] + \mathcal{O}(h^3).
\end{aligned}$$

Comparing this with the Taylor expansion for $x(t_{k+1}) = x(t_k + h)$, we obtain

$$\begin{aligned}
x(t_k + h) &= x(t_k) + hx'(t_k) + \frac{1}{2}h^2 x''(t_k) + \mathcal{O}(h^3) \\
&= x(t_k) + hf(t_k, x_k) + \frac{1}{2}h^2[f_t + f_x x'] + \mathcal{O}(h^3) \\
&= x(t_k) + hf + \frac{1}{2}h^2[f_t + f_x f] + \mathcal{O}(h^3).
\end{aligned}$$

Observing that the first 3 terms are identical, we obtain an estimate on the local truncation error:

$$e_L = |x_{k+1} - x(t_k + h)| = \mathcal{O}(h^3),$$

showing that this is a second order method. $\qquad\square$

Relating Heun's method with the trapezoid integration rule. Integrating the ODE $x' = f(t, x)$ over the integral $t \in [t_k, t_k + h]$, we get

$$\begin{aligned}
x(t_k + h) &= x(t_k) + \int_{t_k}^{t_k+h} x'(t)\, dt \\
&= x(t_k) + \int_{t_k}^{t_k+h} f(t, x(t))\, dt.
\end{aligned}$$

Once $x(t_k) \approx x_k$ is given, then $x_{k+1} \approx x(t_k + h)$ can be computed by suitably approximating the integral. For the Heun's method, we see that

$$K_1 \approx hx'(t_k), \qquad K_2 \approx hx'(t_k + h).$$

Then, the trapezoid rule

$$\int_{t_k}^{t_k+h} x'(t)\, dt \approx \frac{h}{2}\left[x'(t_k) + x'(t_k + h)\right] = \frac{1}{2}(K_1 + K_2),$$

exactly gives Heun's iteration formula.

In general, the Runge-Kutta method of order m takes the form

$$x_{k+1} = x_k + w_1 K_1 + w_2 K_2 + \cdots + w_m K_m,$$

where

$$\begin{cases} K_1 = h \cdot f(t_k, x_k), \\ K_2 = h \cdot f(t_k + a_2 h, x + b_2 K_1), \\ K_3 = h \cdot f(t_k + a_3 h, x + b_3 K_1 + c_3 K_2), \\ \vdots \\ K_m = h \cdot f(t_k + a_m h, x + \sum_{i=1}^{m-1} \phi_i K_i). \end{cases}$$

The parameters w_i, a_i, b_i, ϕ_i are carefully chosen to guarantee the order m. Taylor series is used to verify the order of the method.

Note that the choice of these parameters is NOT unique, since we have many more parameters than constraints.

The classical RK4: This elegant 4-th order method is probably the most commonly used ODE solver.

The classical RK4 algorithm is:

$$x_{k+1} = x_k + \frac{1}{6}\Big[K_1 + 2K_2 + 2K_3 + K_4\Big],$$

where

$$K_1 = h \cdot f(t_k, \ x_k),$$

$$K_2 = h \cdot f(t_k + \frac{1}{2}h, \ x_k + \frac{1}{2}K_1),$$

$$K_3 = h \cdot f(t_k + \frac{1}{2}h, \ x_k + \frac{1}{2}K_2),$$

$$K_4 = h \cdot f(t_k + h, \ x_k + K_3).$$

Relating the RK4 method with Simpson's rule. The integral form of the ODE $x' = f(t, x)$ gives

$$x(t_k + h) = x(t_k) + \int_{t_k}^{t_k+h} x'(t) \, dt = x(t_k) + \int_{t_k}^{t_k+h} f(t, x(t)) \, dt.$$

Once $x(t_k) \approx x_k$ is given, then $x_{k+1} \approx x(t_k + h)$ can be computed by suitably approximating the integral. For the RK4 method, we see that

$$K_1 \approx h x'(t_k),$$
$$K_2 \approx h x'(t_k + h/2),$$
$$K_3 \approx h x'(t_k + h/2),$$
$$K_4 \approx h x'(t_k + h).$$

Then, Simpson's integration rule

$$\int_{t_k}^{t_k+h} x'(t)\,dt \approx \frac{h}{6}\left[x'(t_k) + 4x'(t_k + h/2) + x'(t_k + h)\right]$$

$$= \frac{1}{2}(K_1 + 2K_2 + 2K_3 + K_4)$$

yields the classical RK4 method.

Numerical Simulations. We consider numerical solutions of the ODE

$$x' = -x + e^{-t}, \qquad x(0) = 0.$$

Recall that the exact solution is $x(t) = te^{-t}$.

We will solve this equation with the Heun and the RK4 methods, taking $h = 0.1$, and computing 2 steps. In the following, we thus take $x_0 = 0$, $t_0 = 0$, $t_1 = 0.1$, and $t_2 = 0.2$.

Heun's method. For $k = 0$, we have

$$K_1 = 0.1 * (-x_0 + e^{-t_0}) = 0.1,$$
$$K_2 = 0.1 * (-(x_0 + K_1) + e^{-t_1}) = 0.0804837,$$

so

$$x_1 = x_0 + (K_1 + K_2)/2 = 0.0902419,$$
$$e_1 = 2.4 \cdot 10^{-4}.$$

For $k = 1$, we have

$$K_1 = 0.1 * (-x_1 + e^{-t_1}) = 0.0814596,$$
$$K_2 = 0.1 * (-(x_1 + K_1) + e^{-t_1}) = 0.0647029,$$

so

$$x_2 = x_1 + (K_1 + K_2)/2 = 0.1633231,$$
$$e_2 = 4.2 \cdot 10^{-4}.$$

Classic RK4 method. For $k = 0$, we have

$$K_1 = h * (-x_0 + e^{-0}) = 0.1,$$
$$K_2 = h * (-(x_0 + K_1/2) + e^{-0.05}) = 0.0901229,$$
$$K_3 = h * (-(x_0 + K_2/2) + e^{-0.05}) = 0.0906168,$$
$$K_4 = h * (-(x_0 + K_3) + e^{-0.1}) = 0.0814221$$

so

$$x_1 = x_0 + (K_1 + 2K_2 + 2K_3 + K_4)/6 = 0.0904836,$$
$$e_1 = 2 \cdot 10^{-7}.$$

For $k = 1$, we have

$$K_1 = h * (-x_1 + e^{-0.1}) = 0.0814354,$$
$$K_2 = h * (-(x_1 + K_1/2) + e^{-0.15}) = 0.0729507,$$
$$K_3 = h * (-(x_1 + K_2/2) + e^{-0.15}) = 0.0733749,$$
$$K_4 = h * (-(x_0 + K_3) + e^{-0.2}) = 0.0654872$$

so

$$x_2 = x_1 + (K_1 + 2K_2 + 2K_3 + K_4)/6 = 0.1637459,$$
$$e_2 = 2.6 \cdot 10^{-7}.$$

Clearly, the classic RK4 method gives much better results than both Euler and Heun's methods.

Matlab simulations. The problem can now be solved in Matlab. The numerical solutions with first order forward Euler method and second order Heun's method are given in Figure 9.4, together with the exact solution. One clearly sees that Heun's method gives a better approximation.

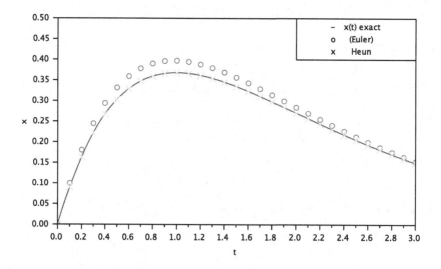

Fig. 9.4   Numerical solutions with Euler and Heun's methods.

The errors with various Runge-Kutta methods are plotted together in Figure 9.5 as a comparison. One sees that the higher order methods perform significantly better than the lower order ones.

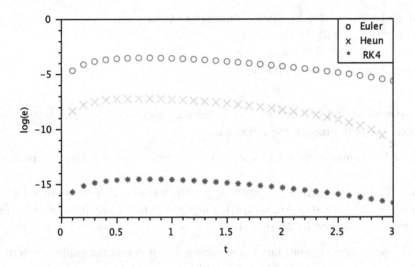

Fig. 9.5   Error comparisons for various RK methods.

9.4   An Adaptive Runge-Kutta-Fehlberg Method

We now consider an adaptive method. In general, for the numerical methods we have considered so far, we have

(1) Smaller time step h gives smaller error;
(2) Higher order methods give better approximations;
(3) With uniform grid, the local error varies at each step, depending on the properties of f and its derivatives.

In order to get a uniform error at each step, we shall allow h to vary at different steps and achieve a better control over the total error. This idea leads to the development of *adaptive methods*, which are based on a careful error analysis.

Main observation: Knowing a solution at time t, we want to compute the solution at $t + h$. If we use two different methods, and one method is more accurate than the other, then we can use the difference between these two solutions as a rough estimate for the local error. This leads to the following basic procedure.

Sketch of the adaptive algorithm.   Fix a tolerance level $\varepsilon > 0$ for the error.

- Compute $x(t + h)$ from $x(t)$ with Method 1, call the solution $x(t+h)$. Method 1 is the less accurate one.
- Compute $x(t + h)$ from $x(t)$ with Method 2, which is a better method that generates a more accurate approximation. Call this solution $\bar{x}(t + h)$;

- Then, $|x(t+h) - \bar{x}(t+h)|$ gives a measure to error:
    if error $\gg \varepsilon$, half the step size;
    if error $\ll \varepsilon$, double the step size;
    if error $\approx \varepsilon$, keep the step size;

Now we need to find two methods with different accuracy. Based on our earlier observations, we may propose the following:

- Method 1: Compute $x(t+h)$ from $x(t)$ with step h, using a specific method, say RK4 method;
- Method 2: Compute $x(t + \frac{1}{2}h)$ from $x(t)$ with step $\frac{1}{2}h$, then compute $x(t+h)$ from $x(t + \frac{1}{2}h)$ with step $\frac{1}{2}h$, with RK4 method. This approximation is better than the one achieved by Method 1.

Here the basic idea is good, but the approach is not computationally efficient.

A better approach, due to Fehlberg, is built upon some higher order RK methods. We start with a 4-th order RK method:

$$x(t+h) = x(t) + \frac{25}{216}K_1 + \frac{1408}{2565}K_3 + \frac{2197}{4104}K_4 - \frac{1}{5}K_5,$$

where

$$K_1 = h \cdot f(t, x),$$
$$K_2 = h \cdot f\left(t + \frac{1}{4}h, \ x + \frac{1}{4}K_1\right),$$
$$K_3 = h \cdot f\left(t + \frac{3}{8}h, \ x + \frac{3}{32}K_1 + \frac{9}{32}K_2\right),$$
$$K_4 = h \cdot f\left(t + \frac{12}{13}h, \ x + \frac{1932}{2197}K_1 - \frac{7200}{2197}K_2 + \frac{7296}{2197}K_3\right),$$
$$K_5 = h \cdot f\left(t + h, \ x + \frac{439}{216}K_1 - 8K_2 + \frac{3680}{513}K_3 - \frac{845}{4104}K_4\right).$$

Adding the additional term

$$K_6 = h \cdot f\left(t + \frac{1}{2}h, \ x - \frac{8}{27}K_1 + 2K_2 - \frac{3544}{2565}K_3 + \frac{1859}{4104}K_4 - \frac{11}{40}K_5\right),$$

we then obtain the 5-th order RK method:

$$\bar{x}(t+h) = x(t) + \frac{16}{135}K_1 + \frac{6656}{12825}K_3 + \frac{28561}{56430}K_4 - \frac{9}{50}K_5 + \frac{2}{55}K_6.$$

Assuming that the 5-th order approximation is very close to the exact solution, we can use the difference $|x(t+h) - \bar{x}(t+h)|$ as an estimate for the error.

This leads to the famous *Adaptive Runge-Kutta-Fehlberg Method*. Below is the pseudo-code for this adaptive RKF45, with a time step controller. This is the actual algorithm used in the commercial Matlab ODE solver of RK4.

Given $t_0, t_f, x_0, h_0, n_{max}, e_{min}, e_{max}, h_{min}, h_{max}$
set $h = h_0, t = t_0, x_0 = x_0, k = 0,$
while $(k < n_{max}$ and $t < t_f)$ do

    if $h < h_{min}$ then $h = h_{min}$,
    else if $h > h_{max}$ then $h = h_{max}$,
    end
    Compute RKF4, RKF5, and $e = |\text{RKF4} - \text{RKF5}|$
    if $(e > e_{max}$ and $h > h_{min})$, then $h = h/2$; (reject the step)
    else –(accept the step)
        $k = k + 1; t = t + h; x_k = \text{RKF5}$;
        If $e < e_{min}$, then $h = 2 * h$; end
    end

end (while)

9.5 Multi-Step Methods

Consider the IVP

$$x' = f(t, x), \qquad x(t_0) = x_0.$$

Let $t_n = t_0 + nh$. If $x(t_n)$ is given, the exact value for $x(t_{n+1})$ would be

$$x(t_{n+1}) = x(t_n) + \int_{t_n}^{t_{n+1}} x'(t)\, dt = x(t_n) + \int_{t_n}^{t_{n+1}} f(t, x(t))\, dt.$$

To use this formula, we need to find a good approximation to the integral

$$\int_{t_n}^{t_{n+1}} f(t, x(t))\, dt.$$

How do we achieve this? One possibility is the following: we approximate f by polynomials, using polynomial interpolation formulas. Then we compute the integral of this polynomial and use it as our approximation.

We now set up the algorithm. Fix an integer k. Assume that $n \geq k$. Given the values

$$(t_{n-k}, x_{n-k}), \quad (t_{n-k+1}, x_{n-k+1}), \quad \cdots \quad (t_n, x_n),$$

we want to compute the approximation x_{n+1}.

Adams-Bashforth (AB) method: The explicit version. Given

$$t_n, t_{n-1}, \cdots, t_{n-k}$$

and

$$x_n, x_{n-1}, \cdots, x_{n-k},$$

we can compute

$$f_n = f(t_n, x_n), \quad f_{n-1} = f(t_{n-1}, x_{n-1}), \quad \cdots, \quad f_{n-k} = f(t_{n-k}, x_{n-k}),$$

and obtain the data set

$$(t_n, f_n), \quad (t_{n-1}, f_{n-1}), \quad \cdots, \quad (t_{n-k}, f_{n-k}).$$

We find a polynomial $P_k(t)$ that interpolates the above data set $(t_i, f_i)_{i=n-k}^{n}$. Recall that there exists a unique such polynomial. Using the Lagrange form, we can write $P_k(t)$ as a linear combination of the cardinal functions $l_i(t)$, namely

$$P_k(t) = f_n l_n(t) + f_{n-1} l_{n-1}(t) + \cdots + f_{n-k} l_{n-k}(t).$$

We then use $P_k(t)$ as an approximation to $f(t, x(t))$, and obtain the iteration formula

$$x_{n+1} = x_n + \int_{t_n}^{t_{n+1}} P_k(s) \, ds.$$

This yields

$$x_{n+1} = x_n + \int_{t_n}^{t_n+h} (f_n l_n(t) + f_{n-1} l_{n-1}(t) + \cdots + f_{n-k} l_{n-k}(t)) \, dt$$

$$= x_n + \int_{t_n}^{t_n+h} f_n l_n(t) \, dt + \int_{t_n}^{t_n+h} f_{n-1} l_{n-1}(t) \, dt + \cdots + \int_{t_n}^{t_n+h} f_{n-k} l_{n-k}(t) \, dt$$

$$= x_n + f_n \int_{t_n}^{t_n+h} l_n(t) \, dt + f_{n-1} \int_{t_n}^{t_n+h} l_{n-1}(t) \, dt + \cdots + f_{n-k} \int_{t_n}^{t_n+h} l_{n-k}(t) \, dt$$

$$= x_n + h \cdot (b_0 f_n + b_1 f_{n-1} + b_2 f_{n-2} + \cdots + b_k f_{n-k}),$$

where $b_0, b_1, \cdots, b_k$ are constants, determined by the integrals of the cardinal functions:

$$b_0 = \frac{1}{h} \int_{t_n}^{t_n+h} l_n(t) \, dt,$$

$$b_1 = \frac{1}{h} \int_{t_n}^{t_n+h} l_{n-1}(t) \, dt,$$

$$\cdots$$

$$b_k = \frac{1}{h} \int_{t_n}^{t_n+h} l_{n-k}(t) \, dt.$$

Example 9.4. If $k = 0$, then we use only one point, i.e., (t_n, f_n), and $P_0(t) = f_n$, which gives

$$x_{n+1} = x_n + h \cdot f(t_n, x_n).$$

We recognize this as the explicit forward Euler's method.

Example 9.5. Now take $k = 1$. In this case the interpolation involves two points. Given x_n, x_{n-1}, we can compute f_n, f_{n-1} as

$$f_n = f(t_n, x_n),$$
$$f_{n-1} = f(t_{n-1}, x_{n-1}).$$

Using a linear interpolation, we obtain

$$x'(s) \approx P_1(s) = f_{n-1} + \frac{f_n - f_{n-1}}{h}(s - t_{n-1}).$$

Then

$$x_{n+1} = x_n + \int_{t_n}^{t_{n+1}} P_1(s)\, ds$$

$$= x_n + \frac{h}{2}(3f_n - f_{n-1}).$$

This is the famous second-order **Adam-Bashforth method**:

$$x_{n+1} = x_n + \frac{h}{2}(3f(t_n, x_n) - f(t_{n-1}, x_{n-1})).$$

One needs two initial values to start the iteration, where $x_0 = x(t_0)$ is given, and x_1 could be computed by either Euler, Heun's or some RK4 method. Since the Adam-Bashforth method is 2nd order, it is better to use Heun's method, to be consistent with the order.

Example 9.6. For $k = 2$, we obtain a third order method:

$$x_{n+1} = x_n + h\left(\frac{23}{12}f_n - \frac{4}{3}f_{n-1} + \frac{5}{12}f_{n-2}\right).$$

For $k = 3$, we have a fourth order method:

$$x_{n+1} = x_n + h\left(\frac{55}{24}f_n - \frac{59}{24}f_{n-1} + \frac{37}{24}f_{n-2} - \frac{3}{8}f_{n-3}\right).$$

Features of these methods:

- Advantages: They are simple, require a minimum number of function evaluations, and are fast to compute.
- Disadvantage: Here we use an interpolating polynomial to approximate a function outside the interval of interpolating points. This is called *extrapolation*, and it gives a bigger error.

Improved version. We now describe an implicit Adams-Bashforth-Moulton (ABM) method.

The main idea here is to avoid using extrapolation, to reduce the interpolation error. We seek a polynomial $P_{k+1}(t)$ that interpolates the data

$$(f_{n+1}, t_{n+1}), \quad (f_n, t_n), \quad \cdots, \quad (f_{n-k}, t_{n-k}),$$

and use it to approximate $f(t, x(t))$, to generate an iteration step

$$x_{n+1} = x_n + \int_{t_n}^{t_{n+1}} P_{k+1}(s)\, ds.$$

Note that we include an additional interpolating point (f_{n+1}, t_{n+1}).

From the previous discussion, using the Lagrange formula for P_{k+1}, we obtain a general formula for each time step:

$$x_{n+1} = x_n + h \cdot (b_{-1} f_{n+1} + b_0 f_n + b_1 f_{n-1} + b_2 f_{n-2} + \cdots + b_k f_{n-k}).$$

Here $b_{-1}, b_0, \cdots, b_k$ are suitable constants, given by the integrals of the cardinal functions.

Important observation: In this method, the value $f_{n+1} = f(t_{n+1}, x_{n+1})$ is unknown! We thus have a non-linear equation to solve for x_{n+1}. For this reason, we call it an **implicit method**.

At each time step, the non-linear equation can be solved by Newton's method or by the secant method. An excellent choice for the initial guess would be the approximate solution provided by the second-order explicit AB method. Newton's method will then converge in 1-2 iterations. In any case, at each step this will require a lot more computations than an explicit method.

Example 9.7. Consider $k = -1, 0, 1, 2, 3$, which corresponds to using 1, 2, 3, 4, 5 points, respectively. Straightforward computations yield

$k = -1: \quad x_{n+1} = x_n + h \cdot f_{n+1}, \qquad$ (implicit backward Euler's method),

$k = 0: \quad x_{n+1} = x_n + \dfrac{h}{2}(f_n + f_{n+1}), \qquad$ (trapezoid rule),

$k = 1: \quad x_{n+1} = x_n + h \cdot \left(\dfrac{5}{12}f_{n+1} + \dfrac{2}{3}f_n - \dfrac{1}{12}f_{n-1}\right),$

$k = 2: \quad x_{n+1} = x_n + h \cdot \left(\dfrac{3}{8}f_{n+1} + \dfrac{19}{24}f_n - \dfrac{5}{24}f_{n-1} + \dfrac{1}{24}f_{n-2}\right),$

$k = 3: \quad x_{n+1} = x_n + h \cdot \left(\dfrac{251}{720}f_{n+1} + \dfrac{646}{720}f_n - \dfrac{264}{720}f_{n-1} + \dfrac{106}{720}f_{n-2} - \dfrac{19}{720}f_{n-3}\right).$

Multi-step method. We now discuss the famous 2-nd order **ABM (Adam-Bashforth-Moulton) method**, which combines the explicit and implicit Adam-Bashforth methods in a smart way. The method has a structure of a predictor step followed by a corrector step. This method is called a **predictor-corrector method**.

A predictor-corrector method:

1. **Predictor step:** Given $x_n, x_{n-1}, f_n, f_{n-1}$, compute the explicit AB solution with $k = 1$:

$$(P) \quad \begin{cases} x_{n+1}^* = x_n + h\left(\dfrac{3}{2}f_n - \dfrac{1}{2}f_{n-1}\right), \\[2mm] f_{n+1}^* = f\left(t_{n+1}, x_{n+1}^*\right). \end{cases}$$

2. **Corrector step:** Take one step of the implicit AB solution with $k = 0$, using the value in step 1 as the value for t_{n+1}:

$$(C) \quad \begin{cases} x_{n+1} = x_n + \dfrac{h}{2}\left(f_{n+1}^* + f_n\right), \\[2mm] f_{n+1} = f(t_{n+1}, x_{n+1}). \end{cases}$$

Numerical Simulations. We consider again the ODE

$$x' = -x + e^{-t}, \qquad x(0) = 0.$$

In order to convince ourselves that the corrector step really reduces the error, we solve the IVP by both the explicit Adam-Bashforth method and the multi-step Adam-Bashforth-Moulton method. Both methods are formally second order, but the first one does not use a corrector step.

The errors are plotted in Figure 9.6. The advantage of the corrector step is apparent in the result.

9.6 A Case Study for a Scalar ODE, Solved in Matlab (optional)

We shall now compare various methods and their approximation errors for a first order ODE. We consider the initial value problem

$$y'(x) = f(x, y) = y + 2x - x^2, \qquad y(0) = -1.$$

The exact solution is

$$y(x) = x^2 - e^x.$$

Note that here we use a different notation, where the unknown function is $y(x)$, and x is the independent variable.

We shall solve the problem with the following methods:

(1) Euler's method,
(2) Heun's method,

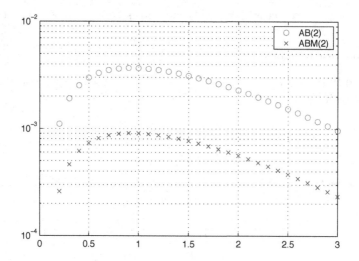

Fig. 9.6   Errors in the numerical solutions with Adam-Bashforth method and the multi-step Adam-Bashforth-Moulton method.

(3) RK4 method,
(4) RKF5 method.

We solve the equation on the interval $x \in [0,1]$, with time step $h = 0.2$, for each method. We provide detailed Matlab codes for each method, and plot all the results for comparison.

9.6.1   *Euler's Method*

Here is the Matlab code for the function `myEuler`:

```
function [y,x]=myEuler(x0,xN,N,y0)
    % input: (x0, xN) = the interval in x
    %        N: total number of intervals
    %        y0: initial value
    % output: (y,x): the solution, y(x).
    h=(xN-x0)/(N); x=[x0:h:xN]; y=zeros(size(x)); y(1)=y0;
    for n=1:1:N,
      k=h*sysfun(x(n),y(n));   y(n+1)=y(n)+k;
    end
end
```

The function $f(x,y)$ is defined in the file `sysfun.m`:

```
function f=sysfun(x,y)
```

```
    f=y+2*x-x*x;
end
```

We also put the exact solution in a function, such as

```
function y=exact(x)
    y=x.*x-exp(x);
end
```

The program should be called by the command:

```
[y,x]=myEuler(0,1,5,-1); plot(x,exact(x),x,y,'.');
```

The plot is shown in Figure 9.7, and the errors are given in Table 9.1.

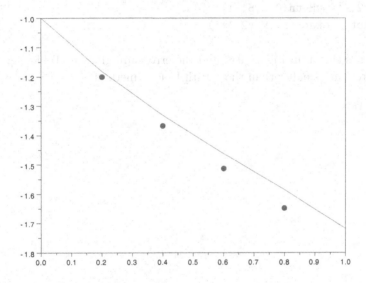

Fig. 9.7   Solution of Euler's method.

Table 9.1   Error in Euler's method

x_n	y_n	$y(x_n)$	$|y_n - y(x_n)|$
0	-1.0000	-1.0000	0
0.2	-1.2000	-1.1814	0.0186
0.4	-1.3680	-1.3318	0.0362
0.6	-1.5136	-1.4621	0.0515
0.8	-1.6483	-1.5855	0.0628
1.0	-1.7860	-1.7183	0.0677

9.6.2 *Heun's Method*

Matlab code for Heun's method, in the file `myHeun.m`, looks like:

```
function [y,x]=myHeun(x0,xN,N,y0)
    h=(xN-x0)/(N); x=[x0:h:xN]; y=zeros(size(x)); y(1)=y0;
    for n=1:1:N,
      fn=sysfun(x(n),y(n));
      k=y(n)+h*fn;
      y(n+1)=y(n)+h/2*(fn+sysfun(x(n+1),k));
    end
end
```

We call the function by:

```
[y2,x]=myHeun(0,1,5,-1);
plot(x,exact(x),x,y2,'*');
```

The plot is shown in Figure 9.8, and the errors are given in Table 9.2. We see that the errors are smaller than those with Euler's method.

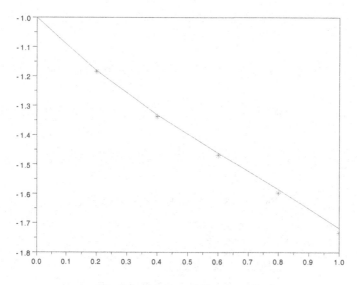

Fig. 9.8   Solution of Heun's method.

9.6.3 *RK4 Method*

Matlab code for RK4 method could be:

Table 9.2   Errors with Heun's method

x_n	y_n	$y(x_n)$	$|y_n - y(x_n)|$
0	−1.0000	−1.0000	0
0.2	−1.1840	−1.1814	0.0026
0.4	−1.3373	−1.3318	0.0055
0.6	−1.4707	−1.4621	0.0086
0.8	−1.5974	−1.5855	0.0119
1.0	−1.7337	−1.7183	0.0154

```
function [y,x] = myRK4(x0,xN,N,y0)
    h=(xN-x0)/(N); x=[x0:h:xN];
    y=zeros(size(x)); y(1)=y0;
    for n=1:1:N,
      k1=h*sysfun(x(n),y(n));
      k2=h*sysfun(x(n)+h/2,y(n)+k1/2);
      k3=h*sysfun(x(n)+h/2,y(n)+k2/2);
      k4=h*sysfun(x(n)+h,y(n)+k3);
      y(n+1)=y(n)+k1/6+k2/3+k3/3+k4/6;
    end
end
```

It could be called by:

```
[y4,x]=myRK4(0,1,5,-1);
plot(x,exact(x),x,y4,'o');
```

The plot is shown in Figure 9.9, and the errors are given in Table 9.3. We see that the errors are really small, much smaller than both Euler's and Heun's methods.

Table 9.3   Errors with RK4 method

x_n	y_n	$y(x_n)$	$|y_n - y(x_n)|$
0	−1.0000	−1.0000	0
0.2	−1.1814	−1.1814	3.9085e-06
0.4	−1.3318	−1.3318	8.0717e-06
0.6	−1.4621	−1.4621	1.2411e-05
0.8	−1.5856	−1.5855	1.6799e-05
1.0	−1.7183	−1.7183	2.1047e-05

9.6.4   RKF5 Method

Matlab code for RKF5 method looks like:

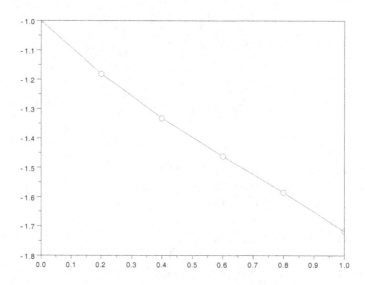

Fig. 9.9   Solution of RK4 method.

```
function [y,x]=myRKF5(x0,xN,N,y0)
  h=(xN-x0)/(N); x=[x0:h:xN]; y=zeros(size(x)); y(1)=y0;
  R=[16/135, 0, 6656/12825, 28561/56430, -9/50, 2/55];
  for n=1:1:N,
    k1=h*sysfun(x(n),y(n));
    k2=h*sysfun(x(n)+h/4,y(n)+k1/4);
    k3=h*sysfun(x(n)+3*h/8, y(n)+3*k1/32+9*k2/32);
    k4=h*sysfun(x(n)+12*h/13, y(n)+1932*k1/2197-7200*k2/2197...
                                                  +7296*k3/2197);
    k5=h*sysfun(x(n)+h,y(n)+439*k1/216-8*k2+3680*k3/513-845*k4/4104);
    k6=h*sysfun(x(n)+h/2,y(n)-8*k1/27+2*k2-3544*k3/2565...
                                          +1859*k4/4104-11*k5/40);
    y(n+1)=y(n)+R(1)*k1+R(2)*k2+R(3)*k3+R(4)*k4+R(5)*k5+R(6)*k6;
  end
end
```

It should be called by:

```
[y5,x]=myRKF5(0,1,5,-1);
plot(x,exact(x),x,y5,'+');
```

The plot is shown in Figure 9.10, and the errors are given in Table 9.4. Again, we see a clear reduction in the errors. This method gives the smallest errors.

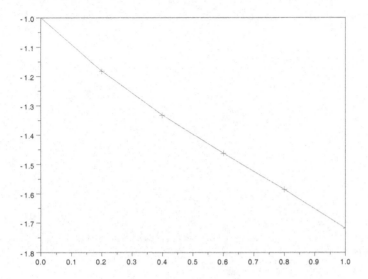

Fig. 9.10   Solution of RKF5 method.

Table 9.4   Errors with RKF5 method

x_n	y_n	$y(x_n)$	$|y_n - y(x_n)|$
0	−1.0000	−1.0000	0
0.2	−1.1814	−1.1814	3.3122e-08
0.4	−1.3318	−1.3318	6.0133e-08
0.6	−1.4621	−1.4621	7.6702e-08
0.8	−1.5855	−1.5855	7.6884e-08
1.0	−1.7183	−1.7183	5.2608e-08

9.6.5   *Comparison*

We now compare these methods. We plot the solutions together in the same graph, listed in Figure 9.11.

The errors for various methods are listed together in Table 9.5 for comparison. It is clear now that the errors become smaller for higher order methods.

Table 9.5   Absolute errors with various methods

x_n	Euler	Heun	RK4	RKF5
0	0	0	0	0
0.2	1.8597e-02	2.5972e-03	3.9085e-06	3.3122e-08
o.4	3.6175e-02	5.4553e-03	8.0717e-06	6.0133e-08
0.6	5.1481e-02	8.5628e-03	1.2411e-05	7.6702e-08
0.8	6.2779e-02	1.1891e-02	1.6799e-05	7.6884e-08
1.0	6.7702e-02	1.5385e-02	2.1047e-05	5.2608e-08

We now run the simulation with a smaller grid size. We let $h = 0.1$, and list the

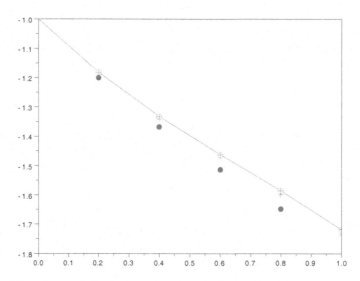

Fig. 9.11   Plot of numerical solutions of various methods. Here the straight line is the exact solution, the dots '•' is the solution by Euler's method, the stars '*' is the solution by Heun's method, the circles 'o' is the solution by RK4 method, and the pluses '+' is the solution by RKF5 method.

errors for various methods in Table 9.6.

Table 9.6   Absolute errors for various methods with $h = 0.1$.

x_n	Euler	Heun	RK4	RKF5
0	0	0	0	0
0.1	4.8291e-03	3.2908e-04	1.2359e-07	5.2611e-10
0.2	9.5972e-03	6.7474e-04	2.5127e-07	1.0099e-09
0.3	1.4241e-02	1.0368e-03	3.8252e-07	1.4368e-09
0.4	1.8685e-02	1.4150e-03	5.1670e-07	1.7892e-09
0.5	2.2840e-02	1.8086e-03	6.5295e-07	2.0470e-09
0.6	2.6598e-02	2.2167e-03	7.9024e-07	2.1863e-09
0.7	2.9836e-02	2.6380e-03	9.2727e-07	2.1792e-09
0.8	3.2407e-02	3.0708e-03	1.0625e-06	1.9935e-09
0.9	3.4139e-02	3.5128e-03	1.1940e-06	1.5918e-09
1.0	3.4835e-02	3.9613e-03	1.3194e-06	9.3046e-10

Let us compare the errors at the point $x_n = 1.0$, for the two grid sizes $h = 0.2$ and $h = 0.1$. We see that, when we half the step size, the error for Euler's method is reduced by a factor $\approx 1/2$, for Heun's method by a factor $\approx 1/4 = 1/(2^2)$, for RK4 by $\approx 1/16 = 1/(2^4)$, and for RKF5 by $\approx 1/32 = 1/(2^5)$. This verifies that Euler's method is 1-st order, Heun's method is 2-nd order, RK4 is 4-th order, while RKF5 is a 5-th order method!

9.7 Numerical Solution of Systems of First Order ODEs

We consider the initial value problem for a system of ODEs

$$\begin{cases} x_1' = f_1(t, x_1, x_2, \cdots, x_n), \\ x_2' = f_2(t, x_1, x_2, \cdots, x_n), \\ \quad \cdots \\ x_n' = f_n(t, x_1, x_2, \cdots, x_n), \end{cases} \qquad \begin{cases} x_1(t_0) = x_1^0, \\ x_2(t_0) = x_2^0, \\ \quad \cdots \\ x_n(t_0) = x_n^0. \end{cases}$$

This can be written in vector notation:

$$\vec{x}' = F(t, \vec{x}), \qquad \vec{x}(t_0) = \vec{x}^0,$$

where $\vec{x} = (x_1, x_2, \cdots, x_n)$ is a vector, and $F = (f_1, f_2, \cdots, f_n)$ is a vector-valued function.

Good news: <u>All</u> numerical methods developed for scalar equation can be adapted to systems! Indeed, the same formulas remain valid with minimal changes, considering x as a vector, and f as a vector-valued function.

For example, the Taylor series method of order m becomes

$$\vec{x}(t+h) = \vec{x} + h\vec{x}' + \frac{1}{2}h^2\vec{x}'' + \cdots + \frac{1}{m!}h^m\vec{x}^{(m)}.$$

Example 9.8. Consider

$$\begin{cases} x_1' = x_1 - x_2 + 2t - t^2 - t^3, \\ x_2' = x_1 + x_2 - 4t^2 + t^3. \end{cases}$$

We shall need the high order derivatives:

$$\begin{cases} x_1'' = x_1' - x_2' + 2 - 2t - 3t^2, \\ x_2'' = x_1' + x_2' - 8t + 3t^2, \end{cases}$$

and

$$\begin{cases} x_1''' = x_1'' - x_2'' - 2 - 6t, \\ x_2''' = x_1'' + x_2'' - 8 + 6t, \end{cases}$$

and so on... The vector computed at step k will be denoted by

$$\vec{x}^k = (x_1^k, x_2^k).$$

The Taylor series method of order 1 (also known as the forward Euler step) becomes

$$\vec{x}^{k+1} = \vec{x}^k + h\left(\vec{x}^k\right)'$$

which we can write out as

$$x_1^{k+1} = x_1^k + h(x_1^k - x_2^k + 2t_k - (t_k)^2 - (t_k)^3),$$
$$x_2^{k+1} = x_2^k + h(x_1^k + x_2^k - 4(t_k)^2 + (t_k)^3).$$

Assume that at the initial time $t_0 = 1$ we are given the initial value $\vec{x}^0 = (1, 1)$, i.e.,

$$x_1^0 = 1, \qquad x_2^0 = 1, \qquad t_0 = 1.$$

Fix the time step $h = 0.1$. The first iteration step will compute an approximation at time $t_1 = 1.1$, namely

$$x_1^1 = x_1^0 + h \left(x_1^0 - x_2^0 + 2t_0 - (t_0)^2 - (t_0)^3 \right) = 1 + 0 = 1,$$
$$x_2^1 = x_2^0 + h \left(x_1^0 + x_2^0 - 4(t_0)^2 + (t_0)^3 \right) = 1 - 0.1 = 0.9.$$

Runge-Kutta methods also take the same form for systems. For example, the classical RK4 becomes:

$$\vec{x}^{k+1} = \vec{x}^k + \frac{1}{6} \left[\vec{K}_1 + 2\vec{K}_2 + 2\vec{K}_3 + \vec{K}_4 \right]$$

where

$$\vec{K}_1 = h \cdot F(t_k, \ \vec{x}^k),$$

$$\vec{K}_2 = h \cdot F(t_k + \frac{1}{2}h, \ \vec{x}^k + \frac{1}{2}\vec{K}_1),$$

$$\vec{K}_3 = h \cdot F(t_k + \frac{1}{2}h, \ \vec{x}^k + \frac{1}{2}\vec{K}_2),$$

$$\vec{K}_4 = h \cdot F(t_k + h, \ \vec{x}^k + \vec{K}_3).$$

When using these formulas, you must keep in mind that here $\vec{x}$ is a vector, instead of a scalar value.

Matlab Codes. The classical RK4 method can be coded as follows, for both scalar ODEs and systems of ODEs.

```
function [t,x] = rk4(f,t0,x0,tend,N)
% f   : Differential equation xp = f(t,x)
% x0  : initial condition
% t0,tend : initial and final time
% N  : number of time steps

h = (tend-t0)/N;
t = [t0:h:tend];
s = length(x0);  % x0 can be a vector
x = zeros(s,N+1);
x(:,1) = x0;

for n = 1:N
   k1 = feval(f,t(n),x(:,n));
```

```
    k2 = feval(f,t(n)+0.5*h,x(:,n)+0.5*h*k1);
    k3 = feval(f,t(n)+0.5*h,x(:,n)+0.5*h*k2);
    k4 = feval(f,t(n)+h,x(:,n)+h*k3);
    x(:,n+1) = x(:,n) + h/6*(k1+2*(k2+k3)+k4);
  end
```

The multi-step second order Adams-Bashforth-Moulton (ABM) method can be coded as follows.

```
function [t,x] = abm2(f,t0,x0,x1,tend,N)
% f  : Differential equation xp = f(t,x)
% x0,x1 : Starting data
% t0,tend : initial time and final time
% N : number of time steps

h = (tend-t0)/N;
t = [t0:h:tend];
s = length(x0);
x = zeros(s,N+1);
x(:,1) = x0;  x(:,2) = x1; % starting data for iterations
fnm1 = feval(f,t(1),x0);
fn = feval(f,t(2),x1);

for n = 2:N
  xs = x(:,n) + 0.5*h*(3*fn-fnm1);  % predictor
  fnp1 = feval(f,t(n+1),xs);  % predictor
  x(:,n+1) = x(:,n)+0.5*h*(fnp1+fn); % corrector
  fnm1 = fn;
  fn = feval(f,t(n),x(:,n)); % corrector
end
```

9.8 Higher Order Equations and Systems

The basic idea for solving higher order equations and systems is to rewrite them as a system of first order equations. In this way, all the methods we learned for systems of first order ODEs can be applied.

An ODE of order n for the scalar function $u(t)$ has the form

$$u^{(n)} = f(t, u, u', u'', \cdots, u^{(n-1)}).$$

At a given time t_0 we consider the initial data

$$u(t_0) = a_0, \qquad u'(t_0) = a_1, \qquad u''(t_0) = a_2, \qquad \cdots \qquad , \qquad u^{(n-1)}(t_0) = a_{n-1}.$$

We show that this problem can be reformulated as the IVP for a system of n first order ODEs. Introduce the variables

$$x_1 = u, \quad x_2 = u', \quad x_3 = u'', \quad \cdots \quad x_n = u^{(n-1)}.$$

Their derivatives are

$$x_1' = u', \quad x_2' = u'', \quad x_3' = u''', \quad \cdots \quad x_n' = u^{(n)}.$$

We thus obtain a system of n first order ODEs

$$\begin{cases} x_1' = x_2, \\ x_2' = x_3, \\ x_3' = x_4, \\ \qquad \vdots \\ x_{n-1}' = x_n, \\ x_n' = f(t, x_1, x_2, \cdots, x_n), \end{cases}$$

with initial data at time $t = t_0$:

$$x_1(t_0) = u(t_0) = a_0,$$
$$x_2(t_0) = u'(t_0) = a_1,$$
$$\cdots$$
$$x_n(t_0) = u^{(n-1)}(t_0) = a_{n-1}.$$

Example 9.9. Consider the second order IVP

$$u'' + 4uu' + t^2 u + t = 0, \qquad u(0) = 1, \quad u'(0) = 2.$$

Introducing the variables

$$x_1 = u(t), \quad x_2 = u'(t),$$

we obtain the system

$$\begin{cases} x_1' = x_2, \\ x_2' = -4x_1 x_2 - t^2 x_1 - t, \end{cases}$$

with initial conditions

$$\begin{cases} x_1(0) = u(0) = 1, \\ x_2(0) = u'(0) = 2. \end{cases}$$

Systems of high-order equations are treated in the same way. We can rewrite each high-order ODE as a system of first order ODEs, and put together all the equations. We illustrate this idea by an example.

Example 9.10. Consider the system of 2-nd order ODEs

$$u'' = u(1 - v') + t,$$
$$v'' = v^2 - u'v' + tu^2,$$

with initial conditions at time $t = 1$:
$$u(1) = 1, \quad u'(1) = 2, \quad v(1) = 3, \quad v'(1) = 4.$$

Introducing the variables
$$x_1 = u, \quad x_2 = u', \quad x_3 = v, \quad x_4 = v',$$

we obtain a system of 4 first order equations:
$$\begin{cases} x_1' = x_2, \\ x_2' = x_1(1 - x_4) + t, \\ x_3' = x_4, \\ x_4' = x_3^2 - x_2 x_4 + t x_1^2, \end{cases}$$

with initial conditions
$$\begin{cases} x_1(1) = u(1) = 1, \\ x_2(1) = u'(1) = 2, \\ x_3(1) = v(1) = 3, \\ x_4(1) = v'(1) = 4. \end{cases}$$

9.9 A Case Study for a System of ODEs by Various Methods (optional)

We consider a test problem
$$y'' + 2y' + 0.75y = 0, \quad y(0) = 3, \quad y'(0) = -2.5.$$

By introducing a vector $\mathbf{y} = (y_1, y_2)$, where
$$y_1 = y, \quad y_2 = y',$$

we get a system
$$\mathbf{y}' = \mathbf{f}(x, \mathbf{y}),$$

where $\mathbf{f} = (f_1, f_2)$ and
$$f_1 = y_2, \qquad f_2 = -2y_2 - 0.75y_1.$$

The exact solution to this problem is
$$y(x) = y_1(x) = 2e^{-0.5x} + e^{-1.5x}$$
$$y'(x) = y_2(x) = -e^{-0.5x} - 1.5e^{-1.5x}.$$

We now compute numerical approximations for this system, using Matlab. First, we define the system in a function:

```
function f=sysfunsys(x,y)
    f=zeros(2,1);
    f(1)=y(2);
    f(2)=-2*y(2)-0.75*y(1);
end
```

The exact solution is also defined in a function:

```
function y=exactsys(x)
  y1=2*exp(-0.5*x)+exp(-1.5*x);
  y2=-exp(-0.5*x)-1.5*exp(-1.5*x);
  y=[y1;y2];
end
```

Matlab code for Euler's method for systems looks like:

```
function [y,x]=sysEuler(x0,xN,N,y0)
    h=(xN-x0)/(N);
    x=[x0:h:xN];
    y=zeros(length(y0),length(x));
    y(:,1)=y0;
    for n=1:1:N,
      k=h*sysfunsys(x(n),y(:,n));
      y(:,n+1)=y(:,n)+k;
    end
end
```

Note the resemblance to the code for the scalar equation. These codes for systems are more general, and they could be used for scalar equations as well.

A Matlab code for Heun's method for systems is :

```
function [y,x]=sysHeun(x0,xN,N,y0)
    h=(xN-x0)/(N);
    x=[x0:h:xN];
    y=zeros(length(y0),length(x));
    y(:,1)=y0;
    for n=1:1:N,
        fn=sysfunsys(x(n),y(:,n));
        k=y(:,n)+h*fn;
        y(:,n+1)=y(:,n)+h/2*(fn+sysfunsys(x(n+1),k));
    end
end
```

And for RK4 method:

```
function [y,x] = sysRK4(x0,xN,N,y0)
    h=(xN-x0)/(N);
```

```
x=[x0:h:xN];
y=zeros(length(y0),length(x));
y(:,1)=y0;
for n=1:1:N,
    k1=h*sysfunsys(x(n),y(:,n));
    k2=h*sysfunsys(x(n)+h/2,y(:,n)+k1/2);
    k3=h*sysfunsys(x(n)+h/2,y(:,n)+k2/2);
    k4=h*sysfunsys(x(n)+h,y(:,n)+k3);
    y(:,n+1)=y(:,n)+k1/6+k2/3+k3/3+k4/6;
end
end
```

Here is a script to call all these functions, and plot the results:

```
x0=0;
x1=1;
N=5;
y0=[3; -2.5];
[y1,x]=sysEuler(x0,x1,N,y0);
[y2,x]=sysHeun(x0,x1,N,y0);
[y4,x]=sysRK4(x0,x1,N,y0);
ex=exactsys(x);

% plots
subplot(2,1,1);
plot(x,ex(1,:),x,y1(1,:),'+',x,y2(1,:),'*', x,y4(1,:),'o' ),
ylabel('y1')
subplot(2,1,2);
plot(x,ex(2,:),x,y1(2,:),'+',x,y2(2,:),'*', x,y4(2,:),'o' ),
ylabel('y2')

%output of error
format short e
[x' ex(1,:)'-y(1,:)' ex(1,:)'-y2(1,:)' ex(1,:)'-y4(1,:)']
[x'  ex(2,:)'-y(2,:)' ex(2,:)'-y2(2,:)' ex(2,:)'-y4(2,:)']
```

The results are shown in Figure 9.12, and the errors for y_1 and y_2 are listed in Table 9.7.

Again, we see that errors are much smaller with higher order methods.

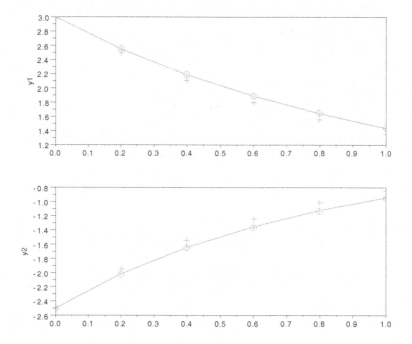

Fig. 9.12   Plots for case study of system of ODEs with various methods. Here: '+' for Euler's method, 'o' for Heun's method, and '*' for RK4 method.

We now define the l^1-norm of a vector $e = (e_1, e_2, \cdots, e_n)$ as

$$\|e\|_{l^1} \doteq \frac{1}{n} \sum_{i=1}^{n} |e_i|.$$

We run the program with various values of h, and we compute the sum of the errors in y_1 and y_2 for each method. The results are given in Table 9.8.

We observe that, when h is halved, errors become smaller for all methods. For Euler's method it is reduced by a factor $\approx 1/2$, for Heun's method it is reduced by $\approx 1/4$, and for RK4 method by $\approx 1/16$. This agrees with the theoretical result that Euler's method is of first order, Heun's method is of second order and RK4 method is of fourth order.

9.10   Stiff Systems

We say that an ODE, or a system of ODEs, is *stiff* if the stability of certain numerical methods requires very small time steps. Stiffness usually occurs when the problem includes some term that varies at a high rate, or the solution is very sensitive to changes in the initial data.

Table 9.7 Errors for y_1 and y_2.

x_n	Euler	Heun	RK4
Absolute errors for y_1			
0	0	0	0
0.2	5.0493e-02	4.5069e-03	1.9443e-05
0.4	7.6273e-02	6.8019e-03	2.8862e-05
0.6	8.7206e-02	7.7228e-03	3.2146e-05
0.8	8.9534e-02	7.8223e-03	3.1841e-05
1.0	8.7141e-02	7.4594e-03	2.9585e-05
Absolute errors for y_2			
0	0	0	0
0.2	6.6065e-02	6.4353e-03	2.9001e-05
0.4	9.6948e-02	9.6143e-03	4.2996e-05
0.6	1.0717e-01	1.0785e-02	4.7816e-05
0.8	1.0586e-01	1.0770e-02	4.7275e-05
1.0	9.8631e-02	1.0099e-02	4.3828e-05

Table 9.8 The l_1 errors for various methods, as the grid size h gets smaller and smaller.

h	Euler	Heun	RK4
1.2500e-02	8.6243e-03	4.7090e-05	7.7373e-10
6.2500e-03	4.3061e-03	1.1724e-05	4.8110e-11
3.1250e-03	2.1516e-03	2.9250e-06	2.9955e-12
1.5625e-03	1.0754e-03	7.3051e-07	1.8903e-13
7.8125e-04	5.3761e-04	1.8253e-07	1.7816e-14

We first illustrate these ideas by a simple scalar equation.

A scalar equation. We first consider a simple scalar equation

$$x' = -ax, \qquad x(0) = 1,$$

where $a > 0$ is a constant, possibly very large. The exact solution is

$$x(t) = e^{-at}.$$

This has an exponential decay. We see that

$$x \to 0 \quad \text{as} \quad t \to +\infty. \tag{9.10.3}$$

Furthermore, the larger the value a, the faster the decay.

We now solve it by explicit forward Euler's method:

$$x_0 = 1, \qquad x_{n+1} = x_n - ahx_n = (1 - ah)x_n, \quad n \geq 1.$$

A straightforward induction argument shows that

$$x_n = (1 - ah)^n x_0 = (1 - ah)^n.$$

We expect that the numerical solution should preserve the important property (9.10.3), i.e.,

$$x_n \to 0, \quad \text{as} \quad n \to +\infty.$$

This will be true if $|1 - ah| < 1$, i.e., if

$$h < \frac{2}{a}. \tag{9.10.4}$$

This gives a restriction on h. Indeed, the step size h must be sufficiently small. The larger the value of a, the smaller h must be, even though the solution is almost zero after a very short time! Such a system, with large a, is called *stiff*.

The condition (9.10.4) is referred to as the *stability condition*. Quite often, to achieve the convergence of an explicit method, a stability condition is required.

To improve the stability, we now use the implicit backward Euler method:

$$x_0 = 1, \qquad x_{n+1} = x_n - ahx_{n+1}, \qquad n \geq 1.$$

This yields

$$x_{n+1} = \frac{1}{1 + ah} x_n.$$

By induction we see that, for all $n \geq 0$,

$$x_n = \left(\frac{1}{1 + ah} \right)^n.$$

Since $ah > 0$, we have

$$0 < \frac{1}{1 + ah} < 1,$$

leading to

$$\lim_{n \to +\infty} x_n = 0,$$

for any choice of $h > 0$. For this reason, we say that the method is **unconditionally stable**.

Remark. (1) Stability conditions can also be introduced for non-linear equations. (2) Usually, implicit methods are unconditionally stable, but they come at a price. Consider the general equation

$$x' = f(t, x), \qquad x(t_0) = x_0,$$

where $f(t, x)$ is non-linear in x. The implicit Euler step is

$$x_{n+1} = x_n + h \cdot f(t_{n+1}, x_{n+1}).$$

To find x_{n+1}, we thus need to solve a non-linear equation. In general, this equation may have one solution, several solutions, or no solution at all. An approximate solution may be obtained using Newton's iteration algorithm, but this can be very time consuming.

Systems of ODEs. Stiffness becomes a harder issue for systems of ODEs. As an example, consider the IVP

$$\begin{cases} x' = -20x - 19y, \\ y' = -19x - 20y, \end{cases} \qquad \begin{cases} x(0) = 2, \\ y(0) = 0. \end{cases}$$

We can rewrite the system using vector notation, as

$$\vec{x}'(t) = A\vec{x}, \qquad A = \begin{pmatrix} -20 & -19 \\ -19 & -20 \end{pmatrix}, \qquad \vec{x} = \begin{pmatrix} x \\ y \end{pmatrix}.$$

For the coefficient matrix A, we have the following eigenvalues and condition number

$$\lambda_1(A) = -1, \qquad \lambda_2(A) = -39, \qquad \text{cond}(A) = 39.$$

Notice that the condition number is rather large.

One can easily check that the exact solution is

$$\begin{cases} x(t) = e^{-39t} + e^{-t}, \\ y(t) = e^{-39t} - e^{-t}. \end{cases}$$

This solution has the following decay property

$$x \to 0, \quad y \to 0 \qquad \text{as} \qquad t \to +\infty. \tag{9.10.5}$$

We notice that:

- There are two components in the solution, e^{-39t} and e^{-t};
- The two eigenvalues give exact the two decay rates in the solution.
- The condition number of A is rather large, indicating two very different rates of decay in the solution. The term e^{-39t} tends to 0 much faster than the term e^{-t};
- For large values of t, the term e^{-39t} becomes insignificant, and the term e^{-t} dominates.
- Therefore, e^{-39t} is called the *transient solution*, and the term e^{-t} is the *steady-state solution*.

Explicit method. We solve the system with forward Euler's method:

$$\begin{cases} x_{n+1} = x_n + h \cdot (-20x_n - 19y_n), \\ y_{n+1} = y_n + h \cdot (-19x_n - 20y_n), \end{cases} \qquad \begin{cases} x_0 = 2, \\ y_0 = 0. \end{cases}$$

One can show by induction that

$$\begin{cases} x_n = (1 - 39h)^n + (1 - h)^n, \\ y_n = (1 - 39h)^n - (1 - h)^n. \end{cases}$$

We must require that the numerical approximation preserves the property (9.10.5), i.e.,

$$x_n \to 0, \qquad y_n \to 0 \qquad \text{as} \qquad n \to +\infty.$$

This gives the conditions

$$|1 - 39h| < 1 \qquad \text{and} \qquad |1 - h| < 1,$$

which implies

$$(1): \ h < \frac{2}{39} \qquad \text{and} \qquad (2): \ h < 2.$$

We see that condition (1) is much stronger than condition (2), therefore it must be satisfied. This is called the *stability condition*.

Condition (1) corresponds to the term e^{-39t}, which is the transient solution and it tends to 0 very quickly as t grows. Unfortunately, time step size is restricted by this transient term even after it has almost no effect on the solution.

Why is this system stiff? Because the condition number is very large, so the system has two components with very different varying rates.

Implicit method. We now propose a more stable method, the implicit Backward Euler method:

$$\begin{cases} x_{n+1} = x_n + h \cdot (-20x_{n+1} - 19y_{n+1}), \\ y_{n+1} = y_n + h \cdot (-19x_{n+1} - 20y_{n+1}), \end{cases} \qquad \begin{cases} x_0 = 2, \\ y_0 = 0. \end{cases}$$

Let

$$A = \begin{pmatrix} -20 & -19 \\ -19 & -20 \end{pmatrix}, \qquad \vec{x} = \begin{pmatrix} x \\ y \end{pmatrix}, \qquad \vec{x}_n = \begin{pmatrix} x_n \\ y_n \end{pmatrix}.$$

We can write

$$\vec{x}_{n+1} = \vec{x}_n + hA \cdot \vec{x}_{n+1},$$

which implies

$$(I - hA)\vec{x}_{n+1} = \vec{x}_n,$$

therefore

$$\vec{x}_{n+1} = (I - hA)^{-1}\vec{x}_n.$$

Taking a vector norm $\| \cdot \|$ of both sides, we obtain

$$\|\vec{x}_{n+1}\| = \left\| (I - hA)^{-1}\vec{x}_n \right\|$$

$$\leq \left\| (I - hA)^{-1} \right\| \cdot \|\vec{x}_n\|.$$

We see that if $\left\| (I - hA)^{-1} \right\| < 1$, then $\vec{x}_n \to 0$ as $n \to +\infty$.

Let us check whether this condition holds, using the l_2 norm:

$$\left\| (I - hA)^{-1} \right\|_2 = \max_i \left| \lambda_i (I - hA)^{-1} \right|$$

$$= \max_i \frac{1}{|(1 - h \cdot \lambda_i(A))|}.$$

Here the two eigenvalues of the symmetric matrix A are

$$\lambda_1(A) = -1, \qquad \lambda_2(A) = -39.$$

Since these are both negative, for $i = 1, 2$ we have $1 - h\lambda_i > 1$. Therefore

$$\left\| (I - hA)^{-1} \right\|_2 < 1,$$

for any choice of the step size $h > 0$. This implicit method is thus *unconditionally stable*.

Main features of the implicit method:

- Advantage: The method is always stable, even for a large step size h. This is particularly important when dealing with a stiff system.
- Disadvantage: At each time step one must solve a system of linear equations

$$(I - hA)\vec{x}_{n+1} = \vec{x}_n .$$

This can be done using Newton's method, or the secant method. In both cases, a longer computing time is required. For this reason, the implicit method is NOT recommended if the system is not stiff.

9.11 Homework Problems for Chapter 9

1. Scalar ODE

Consider the following initial value problem

$$x' = 2x^2 + x - 1, \qquad x(1) = 1.$$

(a). Write out Euler's method for this ODE. Compute the value of $x(1.2)$ using Euler's method, with $h = 0.1$.

(b). Write out the Heun's method for this ODE. Compute the value $x(1.2)$ by Heun's method, with $h = 0.1$.

(c). Write out the classic 4-th order Runge-Kutta method for this ODE. Compute the value $x(1.2)$ by 4-th order Runge-Kutta method, with $h = 0.1$.

(d). Write out the 2-nd order Adams-Bashforth-Moulton method for this ODE. Note that this is a multi-step method. It needs 2 initial values, i.e., x_0 and x_1, to initiate the iterations. For the second initial value x_1, you can use the result obtained in part (b) with Heun's method. Compute the values $x(1.2)$ and $x(1.3)$ using the ABM method.

NB! Do not write Matlab codes for these computations. You may use Matlab as a calculator.

2. Higher Order ODEs

Given the second order equation:

$$x'' - tx = 0, \qquad x(0) = 1, \quad x'(0) = 1,$$

rewrite it as a system of first order equations.

Compute $x(0.1)$ and $x(0.2)$ with 2 time steps using $h = 0.1$, using the following methods:

a) Euler's method,

b) A 2-nd order Runge-Kutta method,

c) A 4-th order Runge-Kutta method,

d) The 2-nd order Adams-Bashforth-Moulton method. Note that this is a multi-step method. For the 2-nd initial value x_1, you can use the solution x_1 from **b)**. For this method, please compute $x(0.2)$ and $x(0.3)$.

NB! Do not write Matlab codes for these computations. You may use Matlab as a fancy calculator.

3. Matlab Solvers: A Case Study in Mechanics

Suppose we have two objects orbiting in space, with masses $1 - \mu$ and μ, rotating around each other. For example, think of the earth and the moon, where the moon moves around the earth at distance 1. (Of course, here both the masses and the distance are normalized.) A third object, which is relatively much smaller and does not affect the motion of the first two, is also orbiting in space. Think of this as a space-ship, or a meteorite.

To simplify the analysis, we assume that all trajectories lie in the same plane. We choose the barycenter of mass as the origin of our coordinate system. Moreover, we adopt a rotating frame of coordinates. In this moving frame, the earth and the moon are always located at the points

$$(y_1^E, y_2^E) = (-\mu, 0), \qquad (y_1^M, y_2^M) = (1 - \mu, 0),$$

and their barycenter is at $(0, 0)$. We denote by $(y_1(t), y_2(t))$ the position of the space ship at time t, w.r.t. this rotating frame of coordinates.

The equations of motion are:

$$y_1'' = y_1 + 2y_2' - \nu \frac{y_1 + \mu}{(d_E)^3} - \mu \frac{y_1 - \nu}{(d_M)^3}$$

$$y_2'' = y_2 - 2y_1' - \nu \frac{y_2}{(d_E)^3} - \mu \frac{y_2}{(d_M)^3}$$

$$d_E = ((y_1 + \mu)^2 + y_2^2)^{1/2}$$

$$d_M = ((y_1 - \nu)^2 + y_2^2)^{1/2}$$

$$\mu = 0.012277471,$$

$$\nu = 1 - \mu.$$

Notice that, in each of the above ODEs, the first two terms account for the Coriolis force, due to the fact that the rotating frame is not an inertial frame. The last two terms account for the gravitational pull of the earth and the moon, respectively. The numbers d_E, d_M measure the distance of the space ship from the earth and from the moon, respectively. Consider the initial conditions

$$y_1(0) = 0.994,$$
$$y_1'(0) = 0,$$
$$y_2(0) = 0,$$
$$y_2'(0) = -2.0015851063790825.$$

For this particular initial data the solution turns out to be periodic. Indeed, at time $t = 17.065211656$ the rotating object goes back to its initial configuration, and the entire motion is repeated over again.

You need to do the following:

(a) Rewrite the system of 2 second order equations as a system of 4 first order ODEs.

(b) Use Matlab function 'ode45' to solve this system of ODEs over one period. Draw the orbit.

(c) Write a program that would solve the system with forward Euler's method, using 24000 uniform time steps.

(d) Write a program that would solve the system with the classical 4-th order Runge-Kutta method, but reduce the step number to 6000.

(e) The total amount of computations in (c) and (d) is about the same. Compare these two solutions. Do you think any of them is acceptable? Explain why.

(f) Find out how many steps 'ode45' used for computing the solution over one period. Can you explain why 'ode45' uses much fewer steps but still gets a much better solution than the other two methods?

(g) Even though the exact solution is periodic, a small error (for example, some perturbation in initial condition) could throw the object out of its orbit. This means that the numerical error, which in other cases might be invisible, could have big consequences in this case. In order to see this effect, you could try to solve the system with 'ode45' over 3 periods. What accuracy (error tolerance) you must require to get an acceptable result? (In this case, "acceptable" means that the plot of the solution looks reasonable to you.)

Hand in whatever you think is relevant. Have fun!

Chapter 10

Numerical Methods for Two-Point Boundary Value Problems

10.1 Introduction

We now consider a second order ODE, having the form

$$y''(x) = f(x, y(x), y'(x)), \qquad y(a) = \alpha, \quad y(b) = \beta. \tag{10.1.1}$$

Here $y(x)$ is the unknown function defined on the interval $a \leq x \leq b$. The values of y (or y') at the two boundary points $x = a$, $x = b$ are given. Such a problem is called a *two-point boundary value problem*. As you might have learned in a sophomore differential equation course, these problems are very different from the initial value problems studied in the previous chapter. For initial value problems, the existence and uniqueness of the solution always hold, provided that the function f is differentiable. However, for two-point boundary value problems, results are very different. Depending on the boundary conditions, the problem might have a unique solution, many solutions, or no solution at all. A detailed discussion on these issues is not within the scope of this book.

Recall that, for initial value problems, the initial conditions are all given at the same time t_0. Here, for two-point boundary value problem, we have boundary conditions at two different points $x = a, x = b$.

An equation such as (10.1.1) arises in many physical models. For example, in the description of an elastic string:

$$y'' = ky + mx(x - L), \qquad y(0) = 0, \quad y(L) = 0.$$

Note that this equation is linear.

One can also have non-linear equations. For example:

$$-(y')^2 - 2b(x)y + 2yy'' = 0, \qquad y(0) = 1, \quad y(1) = a.$$

We shall study two numerical methods for these two-point boundary value problems:

- Shooting method, which is based on the ODE solvers we learned in the previous chapter;
- FDM (Finite Difference Method).

10.2   Shooting Method

The *shooting method*, both as a numerical method and a tool for analysis, is a method for solving a boundary value problem by reducing it to the solution of an initial value problem.

Image you are standing in front of a wall, from a certain distance, and you would like to hit a target on the wall with a rubber gun. We assume that the trajectory of the bullet satisfies a second order ODE. A successful trajectory must satisfy two additional conditions, i.e., the two boundary conditions: start at the point where you stand, and hit the target on the wall. Since the starting point is fixed, you may control one parameter, namely the initial angle of the trajectory. How do you make sure you hit the target? Here is the idea, allowing you several trials. You would make a guess of the initial angle and shoot. You might hit too low. Then you adjust, and try again with a higher angle. You might still miss and hit too high. Then you would choose an intermediate angle, and likely get closer to the target.

The shooting method adopts exactly this procedure. The algorithm goes as follows:

- Solve the equation (10.1.1) as an initial value problem, with two initial conditions given at $x = a$, say $y(a) = \alpha$ and $y'(a) = z$. Here the value α is assigned, but the value z is just a guess.
- Compute the corresponding solution and its value at $x = b$.
- Compare this value with the given boundary condition at $x = b$. Based on this information, you adjust your guess at $x = a$, i.e., make a new guess for the the value $y'(a)$, and iterate if needed, until convergence is observed.

It makes a difference if equation (10.1.1) is linear or non-linear. The linear case is much easier.

10.2.1   *Linear Shooting*

Let us consider the linear two-point boundary value problem in the general form:
$$y''(x) = u(x) + v(x)y(x) + w(x)y'(x), \qquad y(a) = \alpha, \quad y(b) = \beta. \qquad (10.2.2)$$

For linear shooting, no iterations are needed. One only needs to make two different guesses of $y'(a)$. If the two-point boundary value problem has a solution, we shall find it in two steps, as explained below.

- Let $\bar{y}$ solve the same equation as (10.2.2), but with initial conditions:
$$\bar{y}(a) = \alpha, \qquad \bar{y}'(a) = 0. \qquad (10.2.3)$$
  Note that $\bar{y}'(a) = 0$ is the "guess" we make.
- Let $\tilde{y}$ solve the same equation, but with different initial conditions:
$$\tilde{y}(a) = \alpha, \qquad \tilde{y}'(a) = 1. \qquad (10.2.4)$$
  Note that $\tilde{y}'(a) = 1$ is the other "guess" we make.

As we will see later, it doesn't matter what guesses we make here. Any two numbers will work, as long as they are different for $\bar{y}$ and $\tilde{y}$, i.e., $\bar{y}'(a) \neq \tilde{y}'(a)$.

Note that both IVPs in (10.2.3) and (10.2.4) can be written into a system of first order ODEs, and can be solved efficiently by Matlab ODE solvers.

Assume now both equations (10.2.3) and (10.2.4) are now solved on the entire interval $x \in [a, b]$, so that the values $\bar{y}(b)$ and $\tilde{y}(b)$ are computed. Furthermore, we assume $\bar{y}(b) \neq \tilde{y}(b)$ (otherwise the method does not work).[1]

Now let

$$y(x) = \lambda \cdot \bar{y}(x) + (1 - \lambda) \cdot \tilde{y}(x) \qquad (10.2.5)$$

where λ is a constant to be determined, such that $y(x)$ in (10.2.5) becomes the solution for (10.2.2).

We now write a differential equation satisfied by the function y in (10.2.5). Multiplying (10.2.3) by λ and (10.2.4) by $(1 - \lambda)$ and adding the two equations, we obtain

$$
\begin{aligned}
y'' &= \lambda \cdot \bar{y}''(x) + (1 - \lambda) \cdot \tilde{y}''(x) \\
&= \lambda(u + v\bar{y} + w\bar{y}') + (1 - \lambda)(u + v\tilde{y} + w\tilde{y}') \\
&= u + v(\lambda\bar{y} + (1 - \lambda)\tilde{y}) + w(\lambda\bar{y}' + (1 - \lambda)\tilde{y}') \\
&= u + vy + wy'.
\end{aligned}
$$

We see that this y solves the equation (10.2.2) for any choice of λ. Note that the above computation heavily depends on the fact that the equation is linear. If the equation is non-linear, this will not work.

We now check the boundary conditions. At $x = a$, we have

$$y(a) = \lambda\bar{y}(a) + (1 - \lambda)\tilde{y}(a) = \lambda\alpha + (1 - \lambda)\alpha = \alpha.$$

The boundary condition is satisfied for any choice of λ.

At $x = b$, we have

$$y(b) = \lambda\bar{y}(b) + (1 - \lambda)\tilde{y}(b).$$

Since $y(b) = \beta$, this gives us an equation which we can use to find λ, namely

$$\lambda\bar{y}(b) + (1 - \lambda)\tilde{y}(b) = \beta, \qquad \Longrightarrow \qquad \lambda = \frac{\beta - \tilde{y}(b)}{\bar{y}(b) - \tilde{y}(b)}. \qquad (10.2.6)$$

Recall the condition $\bar{y}(b) \neq \tilde{y}(b)$. If $\bar{y}(b) = \tilde{y}(b)$, the value of λ cannot be computed.

Conclusion. The function $y(x)$ in (10.2.5) with λ given in (10.2.6) is the solution of the BVP (Boundary Value Problem) in (10.2.2).

---

[1]If $\bar{y}(b) = \tilde{y}(b)$, then either $\bar{y}(b) = \tilde{y}(b) = \beta$, which indicates multiple solutions for the BVP (Boundary Value Problem), or $\bar{y}(b) = \tilde{y}(b) \neq \beta$, which implies that the BVP has no solution.

Practical issues. One can solve the initial value problems for $\bar{y}$ and $\tilde{y}$ simultaneously. Let

$$y_1 = \bar{y}, \quad y_2 = \bar{y}', \quad y_3 = \tilde{y}, \quad y_4 = \tilde{y}',$$

then

$$\begin{pmatrix} y_1' \\ y_2' \\ y_3' \\ y_4' \end{pmatrix} = \begin{pmatrix} y_2 \\ u + vy_1 + wy_2 \\ y_4 \\ u + vy_3 + wy_4 \end{pmatrix}, \quad \text{IC:} \quad \begin{pmatrix} y_1(a) \\ y_2(a) \\ y_3(a) \\ y_4(a) \end{pmatrix} = \begin{pmatrix} \alpha \\ 0 \\ \alpha \\ 1 \end{pmatrix}.$$

This is a 4×4 system of first order ODEs, which could be solved in Matlab with efficient ODE solvers. The values $\bar{y}(b), \tilde{y}(b)$ will be the last element in the vector y_1, y_3, respectively.

10.2.2 Some Extensions of Linear Shooting

Case 1. Consider the same equation as in (10.2.2), but with different boundary conditions, say

$$y''(x) = u(x) + v(x)y(x) + w(x)y'(x), \quad y(a) = \alpha, \quad y'(b) = \beta. \tag{10.2.7}$$

A similar shooting method can be designed, with minimum adjustment for the boundary condition at $x = b$. The function y in (10.2.5) will satisfy the differential equation, as well as the boundary condition at $x = a$. For the boundary condition at $x = b$, we must require

$$y'(b) = \lambda \bar{y}'(b) + (1 - \lambda)\tilde{y}'(b) = \beta.$$

This gives us a formula to compute the value λ

$$\lambda = \frac{\beta - \tilde{y}'(b)}{\bar{y}'(b) - \tilde{y}'(b)}. \tag{10.2.8}$$

Case 2. Consider a higher order linear equation with boundary conditions

$$y''' = f(x, y, y', y''), \quad y(a) = \alpha, \quad y'(a) = \gamma, \quad y(b) = \beta. \tag{10.2.9}$$

Here $f(x, y, y', y'')$ is an affine function in y, y', y''.

A shooting method can be designed as follows. Let $\bar{y}$ solve the same equation (10.2.9), but with initial conditions:

$$\bar{y}(a) = \alpha, \quad \bar{y}'(a) = \gamma, \quad \bar{y}''(a) = 0. \tag{10.2.10}$$

Let $\tilde{y}$ solve the same equation (10.2.9), but with another initial conditions:

$$\tilde{y}(a) = \alpha, \quad \tilde{y}'(a) = \gamma, \quad \tilde{y}''(a) = 1. \tag{10.2.11}$$

Assume now we solved both equations (10.2.10) and (10.2.11), and the values $\bar{y}(b)$ and $\tilde{y}(b)$ are computed. Let

$$y(x) = \lambda \cdot \bar{y}(x) + (1 - \lambda) \cdot \tilde{y}(x) \tag{10.2.12}$$

where λ is a constant to be determined, such that $y(x)$ in (10.2.12) becomes the solution for (10.2.9).

It is easy to check that y solves the equation in (10.2.9), and satisfies the boundary conditions $y(a) = \alpha, y'(a) = \gamma$, due to the linearity properties. It remains to check the last boundary condition at $x = b$. At $x = b$, we have

$$y(b) = \lambda \bar{y}(b) + (1 - \lambda) \tilde{y}(b) = \beta,$$

which yields the same formula to compute λ, i.e.,

$$\lambda = \frac{\beta - \tilde{y}(b)}{\bar{y}(b) - \tilde{y}(b)}. \tag{10.2.13}$$

Case 3. Consider the same equation as in Case 2, but with different boundary conditions:

$$y''' = f(x, y, y', y''), \qquad y(a) = \beta, \quad y(b) = \alpha, \quad y'(b) = \gamma. \tag{10.2.14}$$

Here $f(x, y, y', y'')$ is an affine function in y, y', y''.

It would be more convenient to treat $x = b$ as the initial point. We can assign initial conditions at $x = b$ and solve the ODE "backward in time". This will leave us one parameter to adjust, namely the value $y''(b)$.

In detail, a shooting method can be designed as follows. Let $\bar{y}$ solve the same equation (10.2.14), but with initial conditions:

$$\bar{y}(b) = \alpha, \quad \bar{y}'(b) = \gamma, \quad \bar{y}''(b) = 0. \tag{10.2.15}$$

Let $\tilde{y}$ solve the same equation (10.2.9), but with another initial conditions:

$$\tilde{y}(b) = \alpha, \quad \tilde{y}'(b) = \gamma, \quad \tilde{y}''(b) = 1. \tag{10.2.16}$$

It is easy to check that y solves the equation in (10.2.14), and satisfies the boundary conditions $y(b) = \alpha, y'(b) = \gamma$, due to the linear properties. It remains to check the last boundary condition at $x = a$. At $x = a$, we have

$$y(a) = \lambda \bar{y}(a) + (1 - \lambda) \tilde{y}(a) = \beta,$$

which give the formula to compute λ, i.e.,

$$\lambda = \frac{\beta - \tilde{y}(a)}{\bar{y}(a) - \tilde{y}(a)}. \tag{10.2.17}$$

10.2.3   *Non-linear Shooting*

We now consider the general form of a second order ODE with boundary conditions

$$y'' = f(x, y, y'), \qquad y(a) = \alpha, \quad y(b) = \beta, \tag{10.2.18}$$

where f is a non-linear function in y or y'.

Since now the ODE is non-linear, the linear method does not work anymore. We must take a new approach, suitable for non-linear problems.

Let $\tilde{y}$ solve the IVP

$$\tilde{y}'' = f(x, \tilde{y}, \tilde{y}'), \qquad \tilde{y}(a) = \alpha, \quad \tilde{y}'(a) = z. \qquad (10.2.19)$$

Note that the condition $\tilde{y}'(a) = z$ is our guess, which is actually the main unknown for the shooting method.

The solution of (10.2.19) depends on the initial condition z. In particular, we can write

$$\tilde{y}(b) \doteq \phi(z),$$

where ϕ is a non-linear function describing how the terminal value $\tilde{y}(b)$ depends on z.

We need to find the value z such that $\phi(z) = \beta$, or equivalently

$$\phi(z) - \beta = 0.$$

Since $\phi(z)$ is a non-linear function, we need to find a root for the above non-linear equation. This was the problem we studied in Chapter 5. Here the non-linear function $\phi(z)$ is not given by an explicit formula, therefore it is very hard (or impossible) to find the derivative $\phi'(z)$. Therefore, Newton's method is not suitable. But we can use the secant method, which does not require the computation of the derivative.

Algorithm. The algorithm goes as follows:

- Choose two initial guesses z_1, z_2, and compute the values

$$\phi_1 = \phi(z_1), \qquad \phi_2 = \phi(z_2).$$

  Note that, to find these values, one needs to solve the initial value problem (10.2.19) with two different initial values: $z = z_1$ and $z = z_2$. Then ϕ_1, ϕ_2 are the values of the two solutions at $x = b$.
- According to the secant method, the next guess z_3 is

$$z_3 = z_2 + (\beta - \phi_2) \cdot \frac{z_2 - z_1}{\phi_2 - \phi_1}.$$

- One can then iterate and find values $z_4, z_5, \cdots$. The procedure is terminated when a stopping criterion is met. For example, when $|\phi(z_n) - \beta| \leq$ tolerance, or when a maximum number of iterations is reached.

Remark: The non-linear shooting could also be used for linear problems. In this case, one iteration will already give us the exact solution. (Can you explain why?)

Extensions. Other types of boundary conditions, as well as higher order non-linear equations, can be handled in a similar way as we did for the linear case. One needs to adopt a secant iteration to adjust the shooting parameter z, as in the above algorithm. Students are encouraged to work out the details on their own.

10.3   Finite Difference Method

We now study the FDM (Finite Difference Method) for the two-point boundary value problem. We consider the linear problem

$$y''(x) = u(x) + v(x)y(x) + w(x)y'(x), \qquad y(a) = \alpha, \quad y(b) = \beta. \qquad (10.3.20)$$

Note that the boundary conditions are given by assigning the values of the unknown function y at the two boundary points. These are called *Dirichlet boundary conditions*.

Discretize the domain: Given $n \geq 1$, we construct a uniform grid by setting

$$h = \frac{b-a}{n}, \qquad x_i = a + ih, \quad i = 0, 1, 2, \cdots, n, \quad x_0 = a, \quad x_n = b.$$

Goal: Find approximations at the discrete grid points: $y_i \approx y(x_i)$.

Tool: We use finite difference approximations to the derivatives:

$$y'(x_i) \approx \frac{y(x_{i+1}) - y(x_{i-1})}{x_{i+1} - x_{i-1}} = \frac{y_{i+1} - y_{i-1}}{2h},$$

$$y''(x_i) \approx \frac{y_{i+1} - 2y_i + y_{i-1}}{h^2}.$$

The discrete equation.   Inserting these two finite difference approximations into the ODE (10.3.20), we obtain

$$\frac{1}{h^2}(y_{i+1} - 2y_i + y_{i-1}) = u_i + v_i y_i + \frac{w_i}{2h}(y_{i+1} - y_{i-1}), \qquad i = 1, 2, \cdots n-1,$$

where we used the notation

$$u_i = u(x_i), \quad v_i = v(x_i), \quad w_i = w(x_i).$$

We can clean up a bit, and get the *discrete equations*

$$-(1 + \frac{h}{2}w_i)y_{i-1} + (2 + h^2 v_i)y_i - (1 - \frac{h}{2}w_i)y_{i+1} = -h^2 u_i, \qquad (10.3.21)$$

for $i = 1, 2, \cdots, n-1$.

Denoting

$$a_i = -(1 + \frac{h}{2}w_i),$$

$$d_i = (2 + h^2 v_i),$$

$$c_i = -(1 - \frac{h}{2}w_i),$$

$$b_i = -h^2 u_i,$$

the discrete equations (10.3.21) can be written in a simpler way

$$a_i y_{i-1} + d_i y_i + c_i y_{i+1} = b_i. \qquad (10.3.22)$$

By the boundary conditions $y_0 = \alpha$ and $y_n = \beta$, the first and last equations in (10.3.22) become

$$d_1 y_1 + c_1 y_2 = b_1 - a_1 \alpha,$$

$$a_{n-1} y_{n-2} + d_{n-1} y_{n-1} = b_{n-1} - c_{n-1}\beta.$$

We see that the equations (10.3.22) lead to a tri-diagonal system of linear equations,

$$\begin{pmatrix} d_1 & c_1 & & & \\ a_2 & d_2 & c_2 & & \\ & \ddots & \ddots & \ddots & \\ & & a_{n-2} & d_{n-2} & c_{n-2} \\ & & & a_{n-1} & d_{n-1} \end{pmatrix} \cdot \begin{pmatrix} y_1 \\ y_2 \\ \vdots \\ y_{n-2} \\ y_{n-1} \end{pmatrix} = \begin{pmatrix} b_1 - a_1\alpha \\ b_2 \\ \vdots \\ b_{n-2} \\ b_{n-1} - c_{n-1}\beta \end{pmatrix}.$$

Using vector notation, this linear system can be written as

$$A\vec{y} = \vec{b}.$$

Discussion. One desirable property to have is that the coefficient matrix A be diagonally dominant. In this case, the linear system can be solved by Gaussian elimination without pivoting. Also, any of the iterative solvers we have learned will converge with any initial guess.

We see that A is strictly diagonally dominant if $|d_i| > |a_i| + |c_i|$, i.e.,

$$|2 + h^2 v_i| > |1 + hw_i/2| + |1 - hw_i/2| \,.$$

This inequality holds if we assume that every term in the absolute value sign above is positive. To achieve this property, we may now require

$$v(x) > 0 \qquad \text{and} \qquad h \le \frac{2}{\max_x |w(x)|}.$$

Example 10.1. Set up the FDM for the boundary value problem

$$y'' = -4(y - x), \qquad y(0) = 0, \quad y(1) = 2.$$

Note that the exact solution is $y(x) = (1/\sin 2)\sin 2x + x$.

Answer. For a fixed $n \ge 1$, we make a uniform grid:

$$h = \frac{1}{n}, \quad x_i = ih, \quad i = 0, 1, 2, \cdots, n.$$

A Central Finite Difference approximation for the second derivative $y''(x_i)$ yields

$$y''(x_i) \approx \frac{1}{h^2}(y_{i-1} - 2y_i + y_{i+1}) = -4y_i + 4x_i.$$

After some cleaning up, we get

$$y_{i-1} - (2 - 4h^2)y_i + y_{i+1} = 4h^2 x_i, \qquad i = 1, 2, \cdots, n - 1,$$

with boundary conditions

$$y_0 = 0, \quad y_n = 2.$$

We thus obtain the tri-diagonal system

$$\begin{pmatrix} -2+4h^2 & 1 & & & \\ 1 & -2+4h^2 & 1 & & \\ & \ddots & \ddots & \ddots & \\ & & 1 & -2+4h^2 & 1 \\ & & & 1 & -2+4h^2 \end{pmatrix} \cdot \begin{pmatrix} y_1 \\ y_2 \\ \vdots \\ y_{n-2} \\ y_{n-1} \end{pmatrix} = \begin{pmatrix} 4h^2 x_1 - 0 \\ 4h^2 x_2 \\ \vdots \\ 4h^2 x_{n-2} \\ 4h^2 x_{n-1} - 2 \end{pmatrix}.$$

The system is "almost" diagonally dominant. It can be solved by Gaussian Elimination efficiently. It could also be solved by any of our iterative methods, and they will all converge.

The graph of the approximate solution with $n = 10$, together with the exact solution, is given in Figure 10.1.

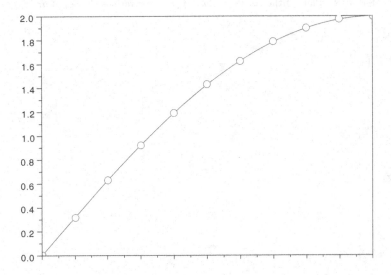

Fig. 10.1 Numerical approximation using the finite difference method, with $n = 10$.

We now plot the error $e_i = y_i - y(x_i)$, for various choices of n, in Figure 10.2. It is clear that the error decreases as n increases.

Finally, we plot $\log(\max(e_i))$ against $\log(h)$ in Figure 10.3. Observe that we get a straight line with slope 2. This indicates a second order method. In general, if a method is of order m, then the error is approximately

$$e \approx Ch^m$$

for some constant C. Taking log on both sides, we get

$$\log(e) = \log(C) + m \log(h),$$

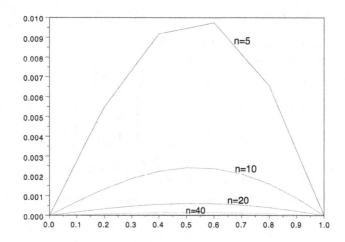

Fig. 10.2   Plots of the error $e_i = |y_i - y(x_i)|$, for various choices of n.

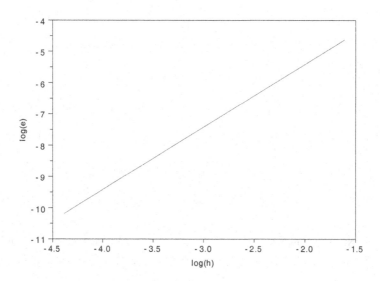

Fig. 10.3   Log-log plot of the error vs. grid size for finite difference approximations with various n.

where m is exactly the slope in the $\log - \log$ plot.

Neumann Boundary conditions. In the previous examples, we assigned the values of the unknown function at the two boundary points $x = a$ and $x = b$. These were called Dirichlet conditions. On the other hand, a *Neumann Boundary* *condition* assigns the first derivative of the unknown function at a boundary point. For example, we consider the Poisson equation in 1 space dimension:

$$u''(x) = f(x), \qquad u'(0) = a, \quad u(1) = b. \qquad (10.3.23)$$

Notice that, at $x = 0$, we are not assigning the value $u(0)$. Instead, we assign the first derivative $u'(0)$. This is a *Neumann boundary condition*.

Consider a uniform grid. Fix an N, let $h = 1/N$ and $x_i = ih$ for $i = 0, 1, 2, \cdots, N$. We also have $u_N = b$ which is a Dirichlet boundary condition.

We now have N unknowns, namely $u_0, u_1, \cdots, u_{N-1}$, and $u_N = b$ is the boundary condition.

We set up the finite difference scheme

$$\frac{u_{i-1} - 2u_i + u_{i+1}}{h^2} = f(x_i),$$

which gives the discrete equations

$$u_{i-1} - 2u_i + u_{i+1} = h^2 f(x_i), \qquad (10.3.24)$$

for $i = 1, 2, \cdots, N - 1$.

Since the central finite difference approximation to $u''(x)$ is second order, we want also to approximate the boundary condition $u'(0) = a$ with a second order finite difference method. The central finite difference for $u'(0)$ is second order, but it requires information at $x = -h$. To handle this, we add an additional grid point outside the domain, $x_{-1} = x_0 - h = -h$. This point is called a *ghost boundary*.

Writing $u_{-1} \approx u(x_{-1})$, we now write out the central finite different for the boundary condition:

$$\frac{u_1 - u_{-1}}{2h} = a, \quad \implies \quad u_{-1} = u_1 - 2ha. \qquad (10.3.25)$$

We also write out (10.3.24) at $i = 0$:

$$u_{-1} - 2u_0 + u_1 = h^2 f(x_0). \qquad (10.3.26)$$

Plugging (10.3.25) into (10.3.26), we get the discrete equation for $i = 0$

$$u_1 - 2ha - 2u_0 + u_1 = h^2 f(x_0),$$

which implies

$$-2u_0 + 2u_1 = h^2 f(x_0) + 2ha. \qquad (10.3.27)$$

The equation for $i = N - 1$ is slightly different due to the boundary condition $u_N = b$. We have

$$u_{N-2} - 2u_{N-1} = h^2 f(x_{N-1}) - b. \qquad (10.3.28)$$

Collecting (10.3.27), (10.3.24) with $i = 1, 2, \cdots, N - 2$, and (10.3.28), we obtain the following tri-diagonal system of linear equations:

$$\begin{pmatrix} -2 & 2 & & & \\ 1 & -2 & 1 & & \\ & \ddots & \ddots & \ddots & \\ & & 1 & -2 & 1 \\ & & & 1 & -2 \end{pmatrix} \cdot \begin{pmatrix} u_0 \\ u_1 \\ \vdots \\ u_{N-2} \\ u_{N-1} \end{pmatrix} = \begin{pmatrix} h^2 f(x_0) + 2ha \\ h^2 f(x_1) \\ \vdots \\ h^2 f(x_{N-2}) \\ h^2 f(x_{N-1}) - b \end{pmatrix}. \qquad (10.3.29)$$

Remark. A Neumann boundary condition at $x = 1$ would be treated in a completely similar way, by adding a ghost boundary point $x_{N+1} = x_N + h$. We omit the details. Students are encouraged to work out the details.

Robin Boundary conditions. For some problem, one might have boundary conditions involving both the unknown and its derivative. These are called *Robin Boundary conditions.*

Consider again the 1D Poisson, but with Robin boundary condition at $x = 0$:

$$u''(x) = f(x), \qquad u'(0) - u(0) = \gamma, \quad u(1) = b. \qquad (10.3.30)$$

Using again the ghost boundary x_{-1}, we can approximate the Robin condition by a second order central finite difference

$$\frac{u_1 - u_{-1}}{2h} - u_0 = \gamma, \quad \Longrightarrow \quad u_{-1} = u_1 - u_0 - 2h\gamma. \qquad (10.3.31)$$

Plugging this into the finite difference scheme (10.3.26), we get the discrete equation for $i = 0$

$$(u_1 - u_0 - 2h\gamma) - 2u_0 + u_1 = h^2 f(x_0),$$

which implies

$$-3u_0 + 2u_1 = h^2 f(x_0) + 2h\gamma. \qquad (10.3.32)$$

The rest of the equations remain unchanged.

We end up with the following tri-diagonal system of linear equations:

$$\begin{pmatrix} -3 & 2 & & & \\ 1 & -2 & 1 & & \\ & \ddots & \ddots & \ddots & \\ & & 1 & -2 & 1 \\ & & & 1 & -2 \end{pmatrix} \cdot \begin{pmatrix} u_0 \\ u_1 \\ \vdots \\ u_{N-2} \\ u_{N-1} \end{pmatrix} = \begin{pmatrix} h^2 f(x_0) + 2h\gamma \\ h^2 f(x_1) \\ \vdots \\ h^2 f(x_{N-2}) \\ h^2 f(x_{N-1}) - b \end{pmatrix}. \qquad (10.3.33)$$

10.4   Homework Problems for Chapter 10

1. *Linear Shooting Method for a Two-point Boundary Value Problem*

Consider the differential equation

$$y'' = y' + 2y + \cos(x), \qquad \text{for} \quad 0 \le x \le \frac{\pi}{2},$$

with boundary conditions

$$y(0) = -0.3, \qquad y\left(\frac{\pi}{2}\right) = -0.1.$$

Show that the exact solution is

$$y(x) = -(\sin(x) + 3\cos(x))/10.$$

Implement the shooting method for this problem in Matlab. Use Matlab solver ode45, with your choice of error tolerance. You can check your answer by comparing it with the exact solution. Plot your solution, and also the error.

2. *Non-linear Shooting Method for a Two-point Boundary Value Problem*

Consider the differential equation

$$y'' = -(y')^2 - y + \ln(x), \qquad 1 \le x \le 2$$

with the boundary conditions

$$y(1) = 0, \qquad y(2) = \ln 2.$$

Show that the exact solution is

$$y(x) = \ln x.$$

Implement the shooting method for this problem in Matlab. Use Matlab solver ode45. Note that this is a non-linear problem, so you need to use a secant iteration. Since the secant iteration converges quickly if the initial guess is good, it is crucial to get a good initial guess. Try the values $z_1 = 1$, $z_2 = 0.5$.

You may choose the tolerance to be 10^{-9}, and maximum number of iterations for the secant method to be 5. Plot the approximate solutions together with the exact solution. Plot also the error.

3. *Finite Difference Method in 1D*

Consider the same equation as in Problem 1. We will now compute approximate solutions with the finite difference method.

(a).   Consider a uniform grid with $h = (b - a)/N$. Set up the finite difference method for the problem. Write out this tri-diagonal system of linear equations for y_i.

(b). Write a Matlab program that computes the approximate solution y_i. You may either use the Matlab solver to solve the linear system, or use the code for tri-diagonal systems (you should find it in a previous homework). Test your program for $N = 10$ and $N = 20$. Plot the approximate solutions together with the exact solution. Plot also the errors.

Chapter 11

Finite Difference Methods for Some PDEs

When the unknown function depends on two or more variables, the derivatives become partial derivatives. The corresponding differential equations are called PDEs (Partial Differential Equations). In this chapter we consider some of the most important second order linear PDEs, and study their finite difference approximations.

These PDEs include:

- The Laplace equation in 2D (i.e., two space dimensions), the Poisson equation in 2D, with Dirichlet and/or Neumann boundary conditions.
- The heat equation in 1D, or other simple parabolic PDEs in 1D, with various boundary conditions.

11.1 Laplace Equation in 2D: Finite Difference Methods

We consider the Laplace equation on the unit square:

$$u_{xx} + u_{yy} = 0, \qquad (0 < x < 1, \quad 0 < y < 1).$$

The Laplace equation belongs to the class of *elliptic PDEs*.

This differential equation must be supplemented with boundary conditions. As a first example, we consider Dirichlet boundary conditions, homogeneous on three sides but non-homogeneous on one side. More precisely:

$$u(0,y) = 0, \quad u(1,y) = 0, \qquad (0 \le x \le 1),$$
$$u(x,0) = 0, \quad u(x,1) = g(x), \qquad (0 \le y \le 1).$$

See Figure 11.1 for an illustration.

The grid. The first step in the discretization is to set up a grid. We shall use a uniform grid in both x and y directions. Choose N, the number of intervals in each direction, and define

$$h = 1/N, \qquad x_i = ih, \qquad y_j = jh, \qquad i,j = 0,1,2,\cdots,N.$$

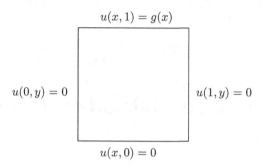

Fig. 11.1 Dirichlet boundary conditions for Laplace equation on a unit square. On 3 sides the unknown $u(x, y)$ is 0, while on the last side the value of u is prescribed.

Our goal is to find an approximate value of the solution at the grid points, i.e., we want to find

$$u_{i,j} \approx u(x_i, y_j), \qquad i, j = 1, 2, \cdots, N - 1.$$

Note that the exact values of $u(x_i, y_j)$ at points on the boundary are already known, thanks to the Dirichlet boundary condition.

Discrete equations. Using the second order central finite difference for the second derivatives, the finite difference approximations to the partial derivatives u_{xx}, u_{yy} at the grid point (x_i, y_j) are

$$u_{xx}(x_i, y_j) \approx \frac{u_{i-1,j} - 2u_{i,j} + u_{i+1,j}}{h^2},$$

$$u_{yy}(x_i, y_j) \approx \frac{u_{i,j-1} - 2u_{i,j} + u_{i,j+1}}{h^2}.$$

Inserting these approximations into the Laplace equation at the point (x_i, y_j), we get

$$\frac{u_{i-1,j} - 2u_{i,j} + u_{i+1,j}}{h^2} + \frac{u_{i,j-1} - 2u_{i,j} + u_{i,j+1}}{h^2} = 0.$$

Multiplying both sides by h^2 and combining the terms with $u_{i,j}$, we obtain the discrete equations

$$\boxed{u_{i-1,j} - 4u_{i,j} + u_{i+1,j} + u_{i,j-1} + u_{i,j+1} = 0, \qquad i, j = 1, 2, \cdots, N - 1. \quad (11.1.1)}$$

Notice that the discrete equation (11.1.1) holds at every grid point (x_i, y_j) in the interior of the domain. In all, we obtain $(N - 1)^2$ linear equations for the unknowns u_{ij}, $i, j = 1, \ldots, N - 1$.

Discrete boundary conditions. The discrete boundary conditions are

$$u_{0,j} = 0, \quad u_{N,j} = 0, \qquad 0 \le j \le N,$$
$$u_{i,0} = 0, \quad u_{i,N} = g_i \doteq g(x_i), \qquad 0 \le i \le N.$$

Computational stencil. Each discrete equation in (11.1.1) involves 5 grid points, which form a **computational stencil**, illustrated in Figure 11.2.

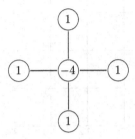

Fig. 11.2 Computational stencil for the discrete Laplace equation.

In the end, we obtain a system of $(N-1)^2$ linear equations in $(N-1)^2$ unknowns, which can be written as

$$A\vec{v} = \vec{b}.$$

Here

$$\vec{v} = (v_1, v_2, \cdots, v_{(N-1)^2})^t$$

is the unknown vector. This vector $\vec{v}$ collects all the unknown variables $u_{i,j}$. Since we have a two-dimensional grid, how do we rearrange these variables $u_{i,j}$ with double indices into a vector $\vec{v}$ with single index? Certainly there are many ways of doing it, and each would lead to an equivalent system of linear equations.

Ordering of unknowns. To form the unknown vector $\vec{v}$ out of the double indexed data u_{ij}, we go through a process called **natural ordering**. It is a rather natural way of reordering, as suggested by the name. We first sweep through the x-direction, then in the y-direction, as follows:

$$\vec{v} = (v_1, v_2, \cdots, v_{(N-1)^2})^T$$
$$= (u_{1,1}, u_{2,1}, \cdots u_{N-1,1}, u_{1,2}, u_{2,2}, \cdots u_{N-1,2}, \cdots, u_{N-1,N-1})^T.$$

Take for example $N = 4$, the total number of unknowns is 9, and the unknown vector $\vec{v}$ formed by natural ordering is

$$v_1 = u_{1,1}, \quad v_2 = u_{2,1}, \quad v_3 = u_{3,1},$$
$$v_4 = u_{1,2}, \quad v_5 = u_{2,2}, \quad v_6 = u_{3,2},$$
$$v_7 = u_{1,3}, \quad v_8 = u_{2,3}, \quad v_9 = u_{3,3}.$$

This ordering is illustrated in Figure 11.3, for $N = 4$.

We can write out the equations at (x_i, y_j), following the same order as the natural ordering, and write out the equations with elements of $\vec{v}$ as the unknowns. We can take advantage of the computational stencil, placing it over each interior grid point. For example, if the center of the stencil is at v_1, then the stencil covers

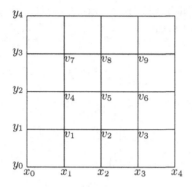

Fig. 11.3   Natural ordering of a 2D grind into a vector, with $N = 4$.

v_2 and v_4, as well as the two boundary points at $(x_1, 0)$ and $(0, y_1)$. Since the values of u at these boundaries are 0, we can remove these terms and get the equation

$$-4v_1 + v_2 + v_4 = 0.$$

We can then go through all the inner points, by placing the center of the stencil at v_k for $k = 1, 2, \cdots, 9$, and write out each equation. Along the upper portion of the boundary, i.e., where $y = y_4 = 1$, we must use the non-homogeneous Dirichlet boundary conditions:

$$u_{i,4} \;=\; u(x_i, y_4) \;=\; g(x_i), \qquad\qquad i = 1, 2, 3.$$

These values will appear on the right-hand sides of the equations centered at v_7, v_8, v_9.

In the end, we obtain the following 9 equations:

$$
\begin{aligned}
-4v_1 + v_2 + v_4 &= 0, \\
v_1 - 4v_2 + v_3 + v_5 &= 0, \\
v_2 - 4v_3 + v_6 &= 0, \\
v_1 - 4v_4 + v_5 + v_7 &= 0, \\
v_2 + v_4 - 4v_5 + v_6 + v_8 &= 0, \\
v_3 + v_5 - 4v_6 + v_9 &= 0, \\
v_4 - 4v_7 + v_8 &= -g(x_1), \\
v_5 + v_7 - 4v_8 + v_9 &= -g(x_2), \\
v_6 + v_8 - 4v_9 &= -g(x_3).
\end{aligned}
$$

We can write them out in the following more organized way:

$$
\begin{array}{lllllll}
-4v_1 \;+v_2 & & +v_4 & & & & = \;\;0, \\
v_1 \;-4v_2 \;+v_3 & & +v_5 & & & & = \;\;0, \\
v_2 \;-4v_3 & & & +v_6 & & & = \;\;0, \\
\hline
v_1 & & -4v_4 \;+v_5 & +v_7 & & & = \;\;0, \\
& v_2 & +v_4 \;-4v_5 \;+v_6 & & +v_8 & & = \;\;0, \\
& & v_3 & +v_5 \;-4v_6 & & +v_9 = & \;\;0, \\
\hline
& & v_4 & & -4v_7 \;+v_8 & = & -g(x_1), \\
& & & v_5 & +v_7 \;-4v_8 \;+v_9 = & -g(x_2), \\
& & & & v_6 & +v_8 \;-4v_9 = & -g(x_3).
\end{array}
$$

This gives us a linear system $A\vec{v} = \vec{b}$, where the coefficient matrix A and load vector $\vec{b}$ are:

$$
A = \left(\begin{array}{ccc|ccc|ccc}
-4 & 1 & 0 & 1 & 0 & 0 & 0 & 0 & 0 \\
1 & -4 & 1 & 0 & 1 & 0 & 0 & 0 & 0 \\
0 & 1 & -4 & 0 & 0 & 1 & 0 & 0 & 0 \\
\hline
1 & 0 & 0 & -4 & 1 & 0 & 1 & 0 & 0 \\
0 & 1 & 0 & 1 & -4 & 1 & 0 & 1 & 0 \\
0 & 0 & 1 & 0 & 1 & -4 & 0 & 0 & 1 \\
\hline
0 & 0 & 0 & 1 & 0 & 0 & -4 & 1 & 0 \\
0 & 0 & 0 & 0 & 1 & 0 & 1 & -4 & 1 \\
0 & 0 & 0 & 0 & 0 & 1 & 0 & 1 & -4
\end{array}\right), \quad
\vec{b} = \left(\begin{array}{c}
0 \\ 0 \\ 0 \\ \hline 0 \\ 0 \\ 0 \\ \hline -g(x_1) \\ -g(x_2) \\ -g(x_3)
\end{array}\right).
$$

Note that the coefficient matrix A is symmetric, sparse and banded. Furthermore, A is diagonally dominant, which is a very desirable property. The most interesting feature is the tridiagonal block-structure of the A matrix. Let

$$
D = \begin{pmatrix} -4 & 1 & 0 \\ -1 & -4 & -1 \\ 0 & 1 & -4 \end{pmatrix}, \quad
I = \begin{pmatrix} 1 & 0 & 0 \\ 0 & 1 & 0 \\ 0 & 0 & 1 \end{pmatrix}, \quad
O = \begin{pmatrix} 0 & 0 & 0 \\ 0 & 0 & 0 \\ 0 & 0 & 0 \end{pmatrix},
$$

then A can be written as a blocked tridiagonal matrix

$$
A = \begin{pmatrix} D & I & O \\ I & D & I \\ O & I & D \end{pmatrix}.
$$

In the general case with an $N \times N$ grid, the same block-structure is preserved, if we form the unknown vector using the natural ordering. Let $D = \text{tridiag}(1, -4, 1)$ be an $(N-1) \times (N-1)$ tridiagonal matrix with -4 on the diagonal and 1 on the sup and sub diagonal, and let $I = \text{diag}(1)$ be an $(N-1) \times (N-1)$ identity matrix. Then, the coefficient matrix A is block-tridiagonal, $(N-1) \times (N-1)$, block structured:

$$
A = \text{tri-diag}(I, D, I) = \begin{pmatrix}
D & I & & & \\
I & D & I & & \\
& \ddots & \ddots & \ddots & \\
& & I & D & I \\
& & & I & D
\end{pmatrix}.
$$

The dimension of A is $(N-1)^2 \times (N-1)^2$.

All boundary conditions are collected in the load vector $\vec{b}$.

11.2   Some Extensions of the Laplace Equation

We now discuss some extensions of the Laplace equation.

Variation 1.   We consider the Laplace equation with non-homogeneous Dirichlet boundary conditions on all 4 sides of the unit squares, given as in Figure 11.4.

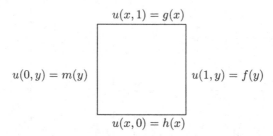

Fig. 11.4   Dirichlet boundary conditions for all 4 sides of a unit square.

Since the boundary conditions only enter the load vector $\vec{b}$, the coefficient matrix A remains unchanged. One may go through the grid again with the computational stencil, and collect all the boundary terms and move them to the load vector.

In the case with $N = 4$, the boundary conditions on the grid are given in Figure 11.5.

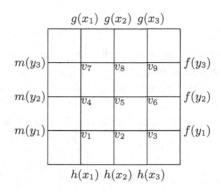

Fig. 11.5   Discrete Dirichlet boundary conditions for $N = 4$.

The load vector b for $N = 4$ is:

$$\vec{b} = \begin{pmatrix} -h(x_1) - m(y_1) \\ -h(x_2) \\ -h(x_3) - f(y_1) \\ \hline -m(y_2) \\ 0 \\ -f(y_2) \\ \hline -g(x_1) - m(y_3) \\ -g(x_2) \\ -g(x_3) - f(y_3) \end{pmatrix}.$$

If $N = 5$ or $N = 6$, the situation is completely similar. The load vectors for these two cases are:

$$N = 5: \quad \vec{b} = \begin{pmatrix} -h(x_1) - m(y_1) \\ -h(x_2) \\ -h(x_3) \\ -h(x_4) - f(y_1) \\ \hline -m(y_2) \\ 0 \\ 0 \\ 0 \\ -f(y_2) \\ \hline -m(y_3) \\ 0 \\ 0 \\ 0 \\ -f(y_3) \\ \hline -g(x_1) - m(y_4) \\ -g(x_2) \\ -g(x_3) \\ -g(x_4) - f(y_4) \end{pmatrix}, \quad N = 6: \quad \vec{b} = \begin{pmatrix} -h(x_1) - m(y_1) \\ -h(x_2) \\ -h(x_3) \\ -h(x_4) \\ -h(x_5) - f(y_1) \\ \hline -m(y_2) \\ 0 \\ 0 \\ 0 \\ -f(y_2) \\ \hline -m(y_3) \\ 0 \\ 0 \\ 0 \\ -f(y_3) \\ \hline -m(y_4) \\ 0 \\ 0 \\ 0 \\ -f(y_4) \\ \hline -g(x_1) - m(y_5) \\ -g(x_2) \\ -g(x_3) \\ -g(x_4) \\ -g(x_5) - f(y_5) \end{pmatrix}.$$

The pattern for a general N should be clear by now.

Variation 2. We now consider the Poisson equation on a unit square

$$u_{xx} + u_{yy} = \phi(x, y)$$

with Dirichlet boundary condition as in Figure 11.4.

The finite difference approximation yields:

$$\frac{u_{i-1,j} - 2u_{i,j} + u_{i+1,j}}{h^2} + \frac{u_{i,j-1} - 2u_{i,j} + u_{i,j+1}}{h^2} = \phi(x_i, y_j) = \phi_{ij}. \quad (11.2.2)$$

This leads to the discrete equations

$$u_{i-1,j} + u_{i+1,j} - 4u_{i,j} + u_{i,j-1} + u_{i,j+1} = h^2\phi_{ij}, \qquad i, j = 1, 2, \cdots, N - 1.$$
$$(11.2.3)$$

We see that the left-hand side of the discrete equation is unchanged. The source term $h^2\phi_{ij}$ will only enter the load vector. Taking for example $N = 4$, we get

$$\vec{b} = \begin{pmatrix} h^2\phi_{1,1} - h(x_1) - m(y_1) \\ h^2\phi_{2,1} - h(x_2) \\ h^2\phi_{3,1} - h(x_3) - f(y_1) \\ \hline h^2\phi_{1,2} - m(y_2) \\ h^2\phi_{2,2} \\ h^2\phi_{3,2} - f(y_2) \\ \hline h^2\phi_{1,3} - g(x_1) - m(y_3) \\ h^2\phi_{2,3} - g(x_2) \\ h^2\phi_{3,3} - g(x_3) - f(y_3) \end{pmatrix}.$$

The students are encouraged to work out the load vector for $N = 5$ and $N = 6$, and eventually find the pattern for a general N.

Variation 3. We now consider a *Neumann boundary condition*, where the derivative of the unknown function is given along part of the boundary. We consider the Laplace equation on a unit square with boundary conditions given in Figure 11.6. We assign Neumann boundary condition on one side of the square, i.e., where $x = 0$.

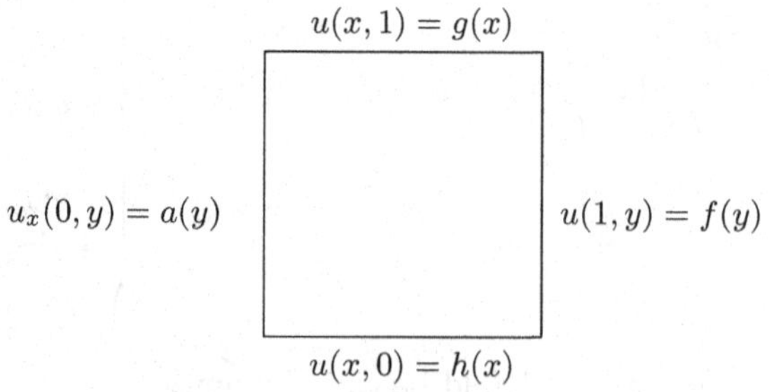

$$u(x, 1) = g(x)$$

$$u_x(0, y) = a(y) \qquad\qquad u(1, y) = f(y)$$

$$u(x, 0) = h(x)$$

Fig. 11.6   Neumann boundary condition on one side of the unit square.

This new boundary condition changes quite a bit the discretization. Since now the values $u_{0,j}$, $j = 1, 2, 3$, are also unknowns, we need to include them into the ordering when we form the vector $\vec{v}$ of the unknowns. The natural ordering in this case is illustrated in Figure 11.7.

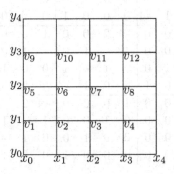

Fig. 11.7 Natural ordering of a 2D grid into a vector with Neumann boundary condition on one side, with $N = 4$.

Since the central finite difference for the second derivatives has second order accuracy, we wish to achieve the same order of accuracy in the Neumann boundary condition. We will use a central finite difference for the first derivative at the boundary. This now involves the point on the left of the boundary $x = 0$, which is outside the domain. For this reason, we add a layer of *ghost boundary points*, i.e., we add the points

$$(x_{-1}, y_j), \qquad \text{for} \quad j = 1, 2, \cdots, N - 1,$$

and denote the discrete values

$$u_{-1,j} \approx u(x_{-1}, y_j), \qquad \text{for} \quad j = 1, 2, \cdots, N - 1.$$

The central finite difference method for the boundary condition at $x = 0$ gives

$$u_x(0, y_j) \approx \frac{u_{1,j} - u_{-1,j}}{2h} = a(y_j) \doteq a_j,$$

which implies

$$u_{-1,j} = u_{1,j} - 2ha_j. \tag{11.2.4}$$

The discrete Laplace equation (11.1.1) at $i = 0$ for any j is

$$u_{-1,j} - 4u_{0,j} + u_{1,j} + u_{0,j-1} + u_{0,j+1} = 0. \tag{11.2.5}$$

Eliminating the ghost value $u_{-1,j}$ from the previous two equations (11.2.4)-(11.2.5), we obtain

$$u_{1,j} - 2ha_j - 4u_{0,j} + u_{1,j} + u_{0,j-1} + u_{0,j+1} = 0,$$

which gives

$$-4u_{0,j} + 2u_{1,j} + u_{0,j-1} + u_{0,j+1} = 2ha_j.$$

This leads to a system of linear equations $A\vec{v} = \vec{b}$. For $N = 4$, the coefficient matrix A becomes

$$
A = \left(
\begin{array}{cccc|cccc|cccc}
-4 & 2 & 0 & 0 & 1 & 0 & 0 & 0 & 0 & 0 & 0 & 0 \\
1 & -4 & 1 & 0 & 0 & 1 & 0 & 0 & 0 & 0 & 0 & 0 \\
0 & 1 & -4 & 1 & 0 & 0 & 1 & 0 & 0 & 0 & 0 & 0 \\
0 & 0 & 1 & -4 & 0 & 0 & 0 & 1 & 0 & 0 & 0 & 0 \\
\hline
1 & 0 & 0 & 0 & -4 & 2 & 0 & 0 & 1 & 0 & 0 & 0 \\
0 & 1 & 0 & 0 & 1 & -4 & 1 & 0 & 0 & 1 & 0 & 0 \\
0 & 0 & 1 & 0 & 0 & 1 & -4 & 1 & 0 & 0 & 1 & 0 \\
0 & 0 & 0 & 1 & 0 & 0 & 1 & -4 & 0 & 0 & 0 & 1 \\
\hline
0 & 0 & 0 & 0 & 1 & 0 & 0 & 0 & -4 & 2 & 0 & 0 \\
0 & 0 & 0 & 0 & 0 & 1 & 0 & 0 & 1 & -4 & 1 & 0 \\
0 & 0 & 0 & 0 & 0 & 0 & 1 & 0 & 0 & 1 & -4 & 1 \\
0 & 0 & 0 & 0 & 0 & 0 & 0 & 1 & 0 & 0 & 1 & -4 \\
\end{array}
\right),
$$

and the load vector $\vec{b}$ is

$$
\vec{b} = \begin{pmatrix}
-h(x_0) + 2ha_1 \\
-h(x_1) \\
-h(x_2) \\
-h(x_3) - f(y_1) \\
2ha_2 \\
0 \\
0 \\
-f(y_2) \\
-g(x_0) + 2ha_3 \\
-g(x_1) \\
-g(x_2) \\
-g(x_3) - f(y_3)
\end{pmatrix}.
$$

We see that A still preserves the block tridiagonal structure. The pattern for A and $\vec{b}$ with larger values of N should now be clear, and we encourage students to work out the details.

11.3 Heat Equation in 1D

We consider a simple heat equation, where the unknown function $u(t, x)$ is the temperature at time t at the point x, in a rod of unit length. This equation has the form

$$u_t = u_{xx}, \qquad 0 \le x \le 1, \qquad t \in [0, T]. \tag{11.3.6}$$

The heat equation belongs to the class of *parabolic PDEs*.

The heat equation must be supplemented with boundary conditions, assigning the values at $x = 0$ and $x = 1$. For example

$$u(t, 0) = 0, \quad u(t, 1) = 0, \qquad \text{for all } t \ge 0, \tag{11.3.7}$$

In addition, one should give an initial condition

$$u(0, x) = f(x) \qquad \text{for all } 0 < x < 1. \tag{11.3.8}$$

These conditions mean that, at the initial time $t = 0$, the temperature distribution on the rod is $u(0, x) = f(x)$. For $t > 0$, the temperature at the two endpoints $x = 0, x = 1$ is kept at the constant value 0.

We want to find the temperature $u(t, x)$ for $t \in [0, T]$ and $x \in [0, 1]$. To solve this problem, we set up a uniform grid on the rectangle $[0, T] \times [0, 1]$.

We divide the space interval $[0, 1]$ into M equal parts, defining

$$\Delta x = \frac{1}{M}, \quad x_j = j\Delta x, \quad j = 0, 1, 2, \cdots, M.$$

Here Δx is the space grid size, and $(M + 1)$ is the total number of points in space.

We then divide the time interval $[0, T]$ into N equal parts, defining

$$t_0 = 0, \quad t_n = n\Delta t, \quad n = 0, 1, 2, \cdots, N, \quad T = t_N = N\Delta t.$$

Here Δt is the time step size, and T is the final computing time.

Our goal is to find approximate to solutions at discrete points, i.e.,

$$u_j^n \approx u(t_n, x_j).$$

We use central finite difference for the double derivative in space, and forward Euler for the time derivative:

$$u_{xx}(t_n, x_j) \approx \frac{u_{j-1}^n - 2u_j^n + u_{j+1}^n}{\Delta x^2}, \qquad u_t(t_n, x_j) \approx \frac{u_j^{n+1} - u_j^n}{\Delta t}.$$

11.3.1 *Forward-Euler Scheme for the Heat Equation*

Plugging the previous finite difference approximations into the PDE

$$u_t(t_n, x_j) = u_{xx}(t_n, x_j),$$

one obtains

$$\frac{u_j^{n+1} - u_j^n}{\Delta t} = \frac{u_{j-1}^n - 2u_j^n + u_{j+1}^n}{\Delta x^2}.$$

Multiplying both sides by Δt, and writing $\gamma = \Delta t/\Delta x^2$, we get the *Forward Euler scheme*:

$$u_j^{n+1} = \gamma u_{j-1}^n + (1 - 2\gamma)u_j^n + \gamma u_{j+1}^n, \tag{11.3.9}$$

with the corresponding discrete initial and boundary conditions

$$u_j^0 = f(x_j), \qquad u_0^n = 0, \quad u_M^n = 0.$$

We see that, in order to compute the value u_j^{n+1}, one only needs to use some values of u at the previous time step n, with a simple formula. Such algorithms are called *explicit methods*.

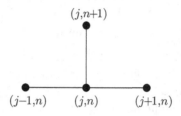

Fig. 11.8   Computational stencil for explicit Euler time step for the heat equation.

The method is first order in time and second order in space, i.e., $\mathcal{O}(\Delta t, \Delta x^2)$. The computational stencil is illustrated in Figure 11.8.

Maximum Principle for the Heat Equation. The exact solution $u(t,x)$ of the heat equation (11.3.6), with boundary conditions (11.3.7) satisfies the following maximum principle:

$$\min_{0 \leq y \leq 1} u(t_1, y) \;\leq\; u(t_2, x) \;\leq\; \max_{0 \leq y \leq 1} u(t_1, y) \qquad \text{for all } x, \qquad (11.3.10)$$

for any $t_2 \geq t_1$.

In particular, (11.3.10) implies

$$\max_{0 \leq x \leq 1} |u(t_2, x)| \;\leq\; \max_{0 \leq x \leq 1} |u(t_1, x)|, \qquad (11.3.11)$$

for any $t_2 \geq t_1$. This means that the maximum value of $|u(t,x)|$ over x is non-increasing in time t.

Discrete Maximum Principle. Since the Maximum Principle is an important feature for the solutions of the heat equation, it is desirable that a discrete version of the maximum principle should hold also for our approximate solutions, namely

$$\max_j |u_j^{n+1}| \leq \max_j |u_j^n|, \qquad \text{for every } n. \qquad (11.3.12)$$

The discrete Maximum Principle is not always satisfied by the numerical solutions. For some methods, certain stability condition must be imposed to ensure (11.3.12).

The Stability Condition. We now provide a sufficient condition for (11.3.12). Assume

$$\gamma \leq \frac{1}{2}, \quad \text{and hence} \quad \Delta t \leq \frac{1}{2}\Delta x^2. \qquad (11.3.13)$$

Then $1 - 2\gamma \geq 0$ and therefore

$$\begin{aligned}
|u_j^{n+1}| &\leq \gamma |u_{j-1}^n| + (1 - 2\gamma) |u_j^n| + \gamma |u_{j+1}^n| \\
&\leq \gamma \max_i |u_i^n| + (1 - 2\gamma) \max_i |u_i^n| + \gamma \max_i |u_i^n| \\
&= \max_i |u_i^n|.
\end{aligned}$$

This means

$$|u_j^{n+1}| \leq \max_i |u_i^n|, \qquad \text{for every } j. \qquad (11.3.14)$$

Since (11.3.14) holds for all j, we can take the maximum over j on the left, and the inequality still holds. Thus we conclude

$$\max_j |u_j^{n+1}| \leq \max_j |u_j^n|, \qquad \text{for all } n.$$

This shows that the discrete Maximum Principle (11.3.12) is true, if we make the assumption (11.3.13). This is called the *stability condition*.

The condition (11.3.13) puts a very strict constraint on the time step size Δt, when space grid size Δx is small. For example, if $\Delta x = 10^{-3}$, then the time step size Δt must be $\Delta t \leq (\frac{1}{2})10^{-6}$, which is extremely small! This forces us to take many time steps!

We will show that this difficulty can be overcome by adopting an implicit time step.

11.3.2 Backward Euler Scheme for the Heat Equation

Instead of a forward Euler, let us now use the backward Euler approximation for the time derivative:

$$u_t(t_{n+1}, x_j) \approx \frac{u_j^{n+1} - u_j^n}{\Delta t}.$$

This yields the backward finite difference scheme:

$$\frac{u_j^{n+1} - u_j^n}{\Delta t} = \frac{u_{j-1}^{n+1} - 2u_j^{n+1} + u_{j+1}^{n+1}}{\Delta x^2}.$$

Writing again $\gamma = \Delta t / \Delta x^2$, we obtain the *Backward Euler Scheme*:

$$-\gamma u_{j-1}^{n+1} + (1 + 2\gamma)u_j^{n+1} - \gamma u_{j+1}^{n+1} = u_j^n, \qquad (11.3.15)$$

with initial and boundary conditions

$$u_j^0 = f(x_j), \qquad u_0^n = 0, \quad u_M^n = 0.$$

Note that scheme (11.3.15) gives a tridiagonal system to solve for the unknown vector at every time step. Such an algorithm is called an *implicit method*. Since the coefficient matrix is strictly diagonally dominant, the system could be solved efficiently by simple Gaussian elimination without pivoting. However, it still takes longer time than the explicit time step in (11.3.9).

This method is first order in time and second order in space, i.e., of $\mathcal{O}(\Delta t, \Delta x^2)$.

The computational stencil is illustrated in Figure 11.9.

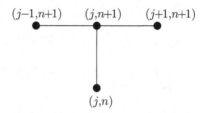

Fig. 11.9 Computational Stencil for implicit Euler time step for the heat equation.

The Discrete Maximum Principle.   We now verify that the discrete Maximum Principle (11.3.12) always holds with this implicit scheme. The discrete equation (11.3.15) can also be written as

$$(1 + 2\gamma)u_j^{n+1} = u_j^n + \gamma u_{j-1}^{n+1} + \gamma u_{j+1}^{n+1}.$$

Taking absolute value on both sides, and using the triangle inequality, we obtain

$$\left|(1 + 2\gamma)u_j^{n+1}\right| \leq \left|u_j^n\right| + \gamma\left|u_{j-1}^{n+1}\right| + \gamma\left|u_{j+1}^{n+1}\right|.$$

Taking the maximum over space index j on the right-hand side, we get

$$(1 + 2\gamma)\left|u_j^{n+1}\right| \leq \left|u_j^n\right| + \gamma\left|u_{j-1}^{n+1}\right| + \gamma\left|u_{j+1}^{n+1}\right|$$
$$\leq \max_i\left|u_i^n\right| + \gamma\max_i\left|u_i^{n+1}\right| + \gamma\max_i\left|u_i^{n+1}\right|$$
$$= \max_i\left|u_i^n\right| + 2\gamma\max_i\left|u_i^{n+1}\right|.$$

Since this inequality is true for all j, it also holds when the left-hand side reaches its maximum value. We conclude

$$(1 + 2\gamma)\max_j\left|u_j^{n+1}\right| \leq \max_j\left|u_j^n\right| + 2\gamma\max_j\left|u_j^{n+1}\right|,$$

which implies

$$\max_j\left|u_j^{n+1}\right| \leq \max_j\left|u_j^n\right|.$$

Notice that the Maximum Principle (11.3.12) is satisfied for any choices of grid size $\Delta t, \Delta x$. For this reason, we say that the scheme is *unconditionally stable*.

Discussion of the implicit scheme:

- Using the implicit scheme, each step requires longer time to compute, but we can take larger time steps Δt. In the end, this method may still save us time!
- However, the algorithm is still of order $\mathcal{O}(\Delta t, \Delta x^2)$, meaning that the error has size $\leq C_1\Delta t + C_2\Delta x^2$. With large time steps Δt, the error is large, and this is not desirable.

There is a way to improve the implicit method, obtaining an algorithm which is of second order both in time and space, and still unconditionally stable.

11.3.3 *Crank-Nicolson Scheme for the Heat Equation*

Both explicit and implicit Euler schemes are first order in time. Since the difference scheme is already of second order w.r.t. the space variable x, it would be desirable to have a method that is also of second order w.r.t. time.

Toward this goal, we use a central finite difference scheme for the time derivative:

$$\frac{u_j^{n+1} - u_j^n}{\Delta t} \approx u_t(t_{n+\frac{1}{2}}, x_j),$$

and approximate the value at the mid-point $t_{n+\frac{1}{2}}$ by the average values at t_n and t_{n+1}, i.e.,

$$u_t(t_{n+\frac{1}{2}}, x_j) \approx \frac{1}{2} u_{xx}(t_n, x_j) + \frac{1}{2} u_{xx}(t_{n+1}, x_j).$$

This gives

$$\frac{u_j^{n+1} - u_j^n}{\Delta t} = \frac{u_{j-1}^n - 2u_j^n + u_{j+1}^n}{2\Delta x^2} + \frac{u_{j-1}^{n+1} - 2u_j^{n+1} + u_{j+1}^{n+1}}{2\Delta x^2}.$$

Writing $r = \Delta t/(2\Delta x^2)$ (note that this is different from the constant γ used in the previous methods!), after some simplifications we obtain the Crank-Nicolson scheme

$$- ru_{j-1}^{n+1} + (1 + 2r)u_j^{n+1} - ru_{j+1}^{n+1} = ru_{j-1}^n + (1 - 2r)u_j^n + ru_{j+1}^n. \qquad (11.3.16)$$

This is supplemented by the initial and boundary conditions

$$u_j^0 = f(x_j), \quad u_0^n = 0, \quad u_M^n = 0.$$

Note that the scheme (11.3.16) yields a tridiagonal system to solve at every time step. Therefore the Crank-Nicolson scheme is an implicit method. The coefficient matrix for the tridiagonal system is symmetric and strictly diagonally dominant.

The computational stencil is shown in Figure 11.10.

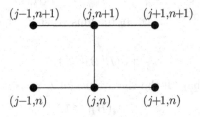

$(j-1,n+1)$ $(j,n+1)$ $(j+1,n+1)$

$(j-1,n)$ (j,n) $(j+1,n)$

Fig. 11.10   The computational stencil for the Crank-Nicolson scheme, for the heat equation.

This method is second order in both time and space, i.e., of $\mathcal{O}(\Delta t^2, \Delta x^2)$.

One can prove that Crank-Nicolson's method is *unconditionally stable*, although the stability is NOT related to the discrete Maximum Principle.

Let v^n denote the column vector of the unknowns, i.e.,

$$v^n = \left(u_1^n, u_2^n, \cdots, u_{M-1}^n\right).$$

Its l_2 norm is given by

$$\|v^n\|_2 = \left(\sum_{i=1}^{M-1} (u_i^n)^2\right)^{1/2}.$$

We recall that the corresponding matrix norm is

$$\|A\|_2 = \max_i |\lambda_i(A)|,$$

where λ_i is the i-th eigenvalue of A, and A is a symmetric matrix.

We have the following Theorem.

Theorem 11.1. *For the Crank-Nicolson scheme, for every n and with any choice of the grid sizes Δt, Δx, the following stability condition holds:*

$$\left\|v^{n+1}\right\|_2 \le \|v^n\|_2.$$

Proof. We begin with rewriting the scheme (11.3.16) in matrix-vector form

$$(I + rA)v^{n+1} = (I - rA)v^n, \tag{11.3.17}$$

where I is the identity matrix of size $(M - 1) \times (M - 1)$, and A is a tridiagonal matrix of the same size, with 2 on the diagonal, and -1 on the lower and upper diagonal.

We see that A is symmetric, diagonally dominant, with positive values along the diagonal. Therefore, A is positive definite, with real and positive eigenvalues.

Thus, the matrix $(I + rA)$ is non-singular and invertible, and (11.3.17) gives

$$v^{n+1} = (I + rA)^{-1}(I - rA)v^n. \tag{11.3.18}$$

Taking the l_2 norm of both sides, we get

$$\begin{aligned}
\left\|v^{n+1}\right\|_2 &= \left\|(I + rA)^{-1}(I - rA)v^n\right\|_2 \\
&\le \left\|(I + rA)^{-1}(I - rA)\right\|_2 \cdot \|v^n\|_2.
\end{aligned}$$

Let $B = (I + rA)^{-1}(I - rA)$. It remains to show that the eigenvalues of this matrix satisfy

$$\left|\lambda_i(B^t B)\right| < 1.$$

This is true provided that all the eigenvalues of B satisfy

$$|\lambda_i(B)| < 1. \tag{11.3.19}$$

From matrix algebra, we have

$$\lambda_i(B) = \frac{1 - r\lambda_i(A)}{1 + r\lambda_i(A)}.$$

Combined with the fact $\lambda_i(A) > 0$ for every i, this implies (11.3.19), completing the proof. $\square$

11.3.4   *The θ-scheme for the Heat Equation*

Let θ be a real number such that $0 \le \theta \le 1$. The following method is called the θ-*scheme*:

$$\frac{u_j^{n+1} - u_j^n}{\Delta t} = (1 - \theta) \cdot \frac{u_{j-1}^n - 2u_j^n + u_{j+1}^n}{2\Delta x^2} + \theta \cdot \frac{u_{j-1}^{n+1} - 2u_j^{n+1} + u_{j+1}^{n+1}}{2\Delta x^2}.$$

We see that the three methods we discussed earlier are all special cases of the θ-scheme. Indeed, we have

- If $\theta = 0$, this gives the explicit forward Euler scheme.
- If $\theta = 1$, this gives the implicit backward Euler scheme.
- If $\theta = \frac{1}{2}$, this gives the Crank-Nicolson scheme.

11.4   Other Forms of the Heat Equation

Neumann Boundary Condition. We now consider the heat equation with Neumann boundary conditions

$$u_x(t, 0) = 0, \qquad u_x(t, 1) = \alpha, \qquad t \ge 0. \tag{11.4.20}$$

Here the condition at $x = 0$ means that one end of the rod is thermally insulated (i.e., heat is not flowing in or out). At $x = 1$ the rod is heated if $\alpha < 0$ and is cooled if $\alpha > 0$.

Let us consider the explicit forward Euler scheme. Setting $\gamma = \Delta t / \Delta x^2$, we obtain the discrete equations

$$u_j^{n+1} = \gamma u_{j-1}^n + (1 - 2\gamma) u_j^n + \gamma u_{j+1}^n, \qquad j = 1, 2, \cdots, M - 1. \tag{11.4.21}$$

In order to handle the boundary conditions at the two end points, we introduce two "ghost" boundary layers, one on each side of the boundary:

$$x_{-1} = -h, \qquad x_{M+1} = x_M + h = b + h.$$

The values of the solution at these points are denoted by

$$u_{-1}^n \approx u(t_n, x_{-1}), \qquad u_{M+1}^n \approx u(t_n, x_{M+1}).$$

Consider the boundary at $j = 0$. Using a second order finite difference, we can approximate the boundary condition as

$$u_x(t_n, x_0) \approx \frac{u_1^n - u_{-1}^n}{2\Delta x} = 0, \qquad (n > 0)$$

which implies

$$u_{-1}^n = u_1^n, \qquad (n > 0). \tag{11.4.22}$$

Assuming that the heat equation holds at $x = 0$, we write the scheme (11.4.21) at $j = 0$,

$$u_0^{n+1} = \gamma u_{-1}^n + (1 - 2\gamma) u_0^n + \gamma u_1^n. \tag{11.4.23}$$

Plugging (11.4.22) into (11.4.23), we can eliminate the term u_{-1}^n, and obtain an equation for $j = 0$:

$$u_0^{n+1} = (1 - 2\gamma)u_0^n + 2\gamma u_1^n. \tag{11.4.24}$$

Next, at the other boundary where $x = x_M = 1$, we again approximate the Neumann boundary condition with a central finite difference,

$$u_x(t_n, x_M) \approx \frac{u_{M+1}^n - u_{M-1}^n}{2\Delta x} = \alpha, \qquad (n > 0).$$

This implies

$$u_{M+1}^n = u_{M-1}^n + 2\Delta x \alpha, \qquad (n > 0). \tag{11.4.25}$$

Assuming now the heat equation holds at $x = x_M = 1$, we write out the scheme (11.4.21) at $j = M$,

$$u_M^{n+1} = \gamma u_{M-1}^n + (1 - 2\gamma)u_M^n + \gamma u_{M+1}^n. \tag{11.4.26}$$

Plugging (11.4.25) into (11.4.26), we eliminate the term u_{M+1}^n, and obtain an equation for $j = M$:

$$u_M^{n+1} = 2\gamma u_{M-1}^n + (1 - 2\gamma)u_M^n + 2\gamma \Delta x \alpha. \tag{11.4.27}$$

Remarks. Students are encouraged to work out the corresponding form of this boundary condition, using the implicit or the Crank-Nicolson scheme, as an exercise.

11.5   Homework Set for Chapter 11

1. FDM for Elliptic Problems in 2D

Consider the Poisson equation for the unknown function $u(x,y)$:

$$u_{xx} + u_{yy} = 1, \qquad 0 < x < 1, \quad 0 < y < 1,$$

with boundary conditions

$$u(0,y) = 0, \quad u(1,y) = y, \qquad 0 \le x \le 1,$$
$$u(x,0) = 0, \quad u(x,1) = x, \qquad 0 \le y \le 1.$$

Consider a uniform grid, choosing $N = 5$ for both the x and y directions, so that

$$h = 1/5, \quad x_i = ih, \quad (i = 0,1,\cdots,5), \qquad y_j = jh \quad (j = 0,1,\cdots,5).$$

Let $u_{i,j} \approx u(x_i, y_j)$ denote the approximate solutions. Using finite different method, set up the system of linear equations for the unknown $u_{i,j}$. (You don't need to solve the system of linear equations.)

2. Heat Equation

Let $u(t,x)$ satisfy the equation

$$u_t(t,x) = 4u_{xx}(t,x) + 1, \qquad 0 < x < 1, \quad t > 0$$

with initial condition

$$u(0,x) = 0, \qquad 0 < x < 1,$$

and boundary conditions

$$u(t,0) = 0, \quad u(t,1) = 0, \qquad t \ge 0.$$

This equation describes the temperature in a rod. The rod initially has a temperature of $0°C$ (zero degree Celsius), and is then heated at a uniform rate 1. However, its two endpoints are kept at the temperature of $0°C$ at all times.

The unknown function $u(t,x)$ describes the temperature in the rod at time $t \ge 0$ at the point $x \in [0,1]$.

(a).   Set up the forward-Euler method.

(b).   Set up the backward-Euler method. Write out the tri-diagonal system one needs to solve at every time step.

Bibliography

Buchanan, J. L. and Turner, P. R. (1992). *Numerical Methods and Analysis* (McGraw-Hill, New York).

Burden, R. L. and Faires, J. D. (2001). *Numerical Analysis* (Brooks/Cole, Pacific Grove, CA, USA, seventh edition).

Kincaid, D. and Cheney, W. (2002). *Numerical Analysis: Mathematics of Scientific Computing.* (Brooks/Cole, Pacific Grove, CA, USA, third edition).

Nelson, D. (eds.) (2008). *The Penguin Dictionary of Mathematics* (Penguin, London, fourth edition).

Pozrikidis, C. (1998). *Numerical Computation in Science and Engineering* (Oxford University Press, New York).

Shen, W. (2015). *Live recordings of lectures and video recordings of topics.* `http://www.youtube.com/wenshenpsu`. (Personal Youtube Channel).

Sli, E. and Mayers, D. F. (2003). *An Introduction to Numerical Analysis* (Cambridge University Press, Cambridge, UK).

Turner, P. (2001). *Guide to Scientific Computing* (CRC Press).

MathWorks (2015). *Matlab online tutorial.* `http://www.mathworks.com/academia/student_center/tutorials/launchpad.html`. (MathWorks online resource)

MathWorks (2015). *Getting Started with Matlab.* `http://www.mathworks.com/help/pdf_doc/matlab/getstart.pdf`. (MathWorks online resource)

Index

Printed in the United States
By Bookmasters